Governing China in the 21st Century

Since 1978, China's political and social systems have transformed significantly to accommodate the world's largest population and second largest economy. These changes have grown more complex and challenging as China deals with modernization, globalization, and informatization. The unique path of sociopolitical development of China hardly fits within any existing frame of reference. The number of scientific explorations of China's political and social development, as well as contributions to international literature from Chinese scholars living and researching in Mainland China, has been growing fast. This series publishes research by Chinese and international scholars on China's politics, diplomacy, public affairs, and social and economic issues for the international academic community.

More information about this series at
http://www.palgrave.com/gp/series/15023

Dan Guttman · Yijia Jing · Oran R. Young
Editors

Non-state Actors in China and Global Environmental Governance

palgrave
macmillan

Editors
Dan Guttman
Tianjin University Law School
Tianjin, China

Fudan University Institute
for Global Public Policy
Shanghai, China

New York University
US Asia Law Institute
New York, USA

Yijia Jing
School of International Relations
and Public Affairs
Fudan University
Shanghai, China

Oran R. Young
Bren School of Environmental Science
and Management
University of California Santa Barb
Santa Barbara, CA, USA

ISSN 2730-6968 ISSN 2730-6976 (electronic)
Governing China in the 21st Century
ISBN 978-981-33-6593-3 ISBN 978-981-33-6594-0 (eBook)
https://doi.org/10.1007/978-981-33-6594-0

Image credit: Xuanyu Han

This Palgrave Macmillan imprint is published by the registered company Springer Nature Singapore Pte Ltd.
The registered company address is: 152 Beach Road, #21-01/04 Gateway East, Singapore 189721, Singapore

Contents

Notes on Contributors

Barbara Bramble Vice President, International Conservation and Corporate Strategies, National Wildlife Federation. Barbara works with the private sector to sever the link between deforestation and agricultural production; she helps global brands and retailers to avoid purchasing agricultural and forest commodities that originate from recently cleared tropical forests and other carbon-rich lands, and to implement voluntary certification standards for sustainable products. For over three decades, she has directed NWF's advocacy to improve U.S. international environmental policy with regard to climate change and forest conservation, and the social and environmental policies of multi-lateral financial institutions. She is the Chair of the Board of the Forest Stewardship Council, the preeminent eco-label for wood and paper products from sustainably managed forests, and formerly chaired the Board of Directors of the Roundtable on Sustainable Biomaterials, which is the equivalent organization for biofuels and bio-based products. Before joining NWF, she served as legal advisor to the White House Council on Environmental Quality and as an environmental lawyer in private practice.

Maoliang Bu is currently an Associate Professor at Nanjing University and Visiting Professor at Ivey Business School. He has published in leading academic journals, among others, *NATURE Sustainability*, *Journal of International Business Studies* (JIBS), *Strategic Management Journal* (SMJ), and *Journal of Comparative Economics*. He has served as an editor for *Journal of Business Ethics*, *Asia Pacific Journal of Management* (APJM), *Journal of International Management* (JIM), and other journals.

Dr. Bu is also an Adjunct Professor at Hopkins-Nanjing Center (Johns Hopkins University, School of Advanced International Studies) and an Affiliated Researcher at Copenhagen Business School. Previously, he has worked as a Postdoc/Visiting Professor at several universities such as University of Goettingen, University of Wuerzburg, University of Gothenburg, and University of Groningen. He has been awarded a prestigious fellowship from the Alexander von Humboldt Foundation.

Ruoyu Chen is a recent graduate from New York University's Stern School of Business majoring in business and political economy and will soon begin his graduate studies at Yale University's International and Development Economics program. His main academic and professional interests lie in China-Africa cooperation and its economic, social, and environmental implications.

Ruoyu co-founded Africa Info & Consulting, an information and research platform aimed at providing in-depth analyses about Africa's political and economic situations, industry outlooks, and development-related issues for Chinese audiences. He authors a quarterly South Africa economic and politics report and co-authors special columns on information and communication technology, renewable energy, and climate change.

Richard ("Tad") Ferris is Senior Counsel at the Institute for Governance & Sustainable Development (IGSD), in Washington, DC, USA, and Paris, France. Ferris brings decades of private law practice experience advising enterprises, multilateral institutions, and nonprofit organizations on Chinese and international environmental, health, safety, and climate law, including product standards and supplier compliance. Ferris was also General Counsel to and Founder of the China Environment, Health and Safety (EHS) Roundtable, an organization dedicated to advancing compliance leadership in China and elsewhere around the world. Ferris began his professional career as Asia Program Manager at the Center for International Environmental Law, working with Durwood Zaelke, IGSD's President. Ferris received J.D. and LL.M. degrees from Duke University School of Law. He also holds a Master's Degree in East Asian Studies (with distinction) and a Bachelor's Degree in Political Science and International Relations. Ferris speaks multiple languages, including Mandarin Chinese, Japanese, and French.

Dan Guttman is teacher, lawyer and has been public servant. He directed US Senate investigations of government management, a presidential bioethics committee, and was Presidentially appointed Occupational Safety and Health Review Commission Commissioner, and UNDP and EU foreign expert advisor on China environmental law. He was partner in a law firm that pioneered in applying antimonopoly law to the energy industry and is of counsel to Guttman, Buschner, and Brooks, whose lawyers represent whistleblowers and have helped government recover billions of dollars from energy and healthcare companies and military contractors. Following 2004–2006 years as Fulbright Scholar in China, he has worked with colleagues at Fudan, Peking, Tsinghua, Tianjin, Nanjing, and Shanghai Jiao Tong universities teaching and developing programs in law, public policy, environment, and international relations. He is Professor at Tianjin University Law School, Adjunct Professor at Fudan IGPP, and New York University US Asia Law Institute Fellow. He is Fellow of the US National Academy of Public Administration, shared in investigative journalism awards, and authored/co-authored many books and articles, including *The Shadow Government*, a seminal study of US government by third party.

Tao Hu is the President of Lakestone Institute for Sustainable Development (LISD), Senior Adviser to WWF China, Chairman of Board of Professional Association for China's Environment (PACE), and Vice President of the Union of Chinese American Professional Organizations (UCAPO). Prior he was WWF-US's China Program Managing Director, Senior Fellow of World Resources Institute (WRI), and Senior Fellow of Policy Research Center of Ministry of Environmental Protection (MEP), China. He was a Visiting Professor at Center for Asia and Pacific Studies

(CAPS) of University of Oregon, USA, and a Visiting Professor of Environmental School, Beijing Normal University. He was also the Chief Expert of Trade and Environment Expert Group for WTO New Round Negotiation, Ministry of Environmental Protection (MEP) of China. He severed as a member of Lead Expert Group China Council for International Cooperation on Environment and Development (CCICED) during 2001–2007.

Dr. Hu's research covers Environmental Economics, Policy and Governance, Trade, Investment and Environment, Globalization and B&R, Co-control of Air Pollutants and GHGs, 3E (Energy, Environment and Economics) Policy, etc.

Dr. Yijia Jing is a Chang Jiang Scholar, Dean of the Institute for Global Public Policy, and Professor of the School of International Relations and Public Affairs, Fudan University. He conducts research on privatization, governance, and collaborative service delivery. He is editor-in-chief of "Fudan Public Administration Review" and co-editor of *International Public Management Journal*. He is the founding co-editor of the Palgrave book series, *Governing China in the 21st Century*, and the founding journal editor of *Global Public Policy and Governance.*

Yifei Li is Assistant Professor of Environmental Studies at NYU Shanghai and Global Network Assistant Professor at NYU. His research concerns both the macro-level implications of Chinese environmental governance for state-society relations, marginalized populations, and global ecological sustainability, and the micro-level bureaucratic processes of China's state interventions into the environmental realm. He has received research support from the United States National Science Foundation, the Rachel Carson Center for Environment and Society, the University of Chicago Center in Beijing, and the China Times Cultural Foundation, among other extramural sources. He is coauthor (with Judith Shapiro) of *China Goes Green: Coercive Environmentalism for a Troubled Planet* (2020). His recent work appears in *Current Sociology, International Journal of Urban and Regional Research, Environmental Sociology*, and other scholarly outlets. He received his Ph.D. in Sociology from the University of Wisconsin-Madison.

Dr. Lingxuan Liu is a Lecturer of Sustainability in Pentland Centre, Lancaster University, and a practitioner (from NGOs & industries) on sustainable supply chain management and urban environmental governance. He completed Ph.D. study at Nanjing University, China, on environmental management and policy, during which he managed several research programs about eco-industrial parks and regional environmental policies. He also spent two years at University of California, Santa Barbara, for an exchange program supported by China Scholarship Council. Lingxuan's work applies longitudinal observations and qualitative analysis to various organizations and the institutional arrangement across different nations that can generate or prohibit sustainability solutions in both the public and private sectors. Such observation and analysis provided interdisciplinary insights across public policy as well as management studies and environmental science. He also collaborates a lot with quantitative researchers using longitudinal data such as national and provincial environmental statistics, or the business and environmental performance of manufacturing companies.

Kate Logan served as international outreach director from 2015 to 2018 at the Institute of Public & Environmental Affairs (IPE), where she focused on integrating transparency and stakeholder participation into corporations' supply chain environmental management systems. Prior to joining IPE, she researched the implementation of environmental public interest litigation in China as a Princeton-in-Asia Fellow on the China Environmental Law & Governance project of the Natural Resources Defense Council. Kate is a co-founder of the Beijing Energy Network's podcast, *Environment China*, and is an avid distance and trail runner. She recently completed her master's at Yale University's School of the Environment, where she was named an EDF Climate Corps Fellow, and also holds a bachelor's *summa cum laude* from Middlebury College with highest honors in East Asian Studies.

Lydia J. Price is Professor of Marketing and Responsible Business at China Europe International Business School (CEIBS) in Shanghai, China. She has served as Chair of the Marketing Department, Associate Dean of the M.B.A. Program, and Head of International Accreditation at CEIBS and has contributed significantly to the school's internationalization and commitment to responsible leadership. Prior to joining CEIBS, Professor Price was in the faculties of Hong Kong University of Science and Technology, and INSEAD, France. She also has been a Visiting Professor at New York University, Beijing University, and Catholic University of Lisbon, Portugal. Professor Price holds a Ph.D. in Marketing from Columbia University, an M.S. in Marketing Research from University of Cincinnati, and a B.S. in Marketing from Miami University.

Michael Spencer was the founding Co-Chair of the Edinburgh-based Alliance for Water Stewardship (AWS) and founding CEO of AWS in the Asia-Pacific. He had previously been the founding CEO of the Forest Stewardship Council (FSC) in Australia after working as Head of Marketing and Communication at FSC International in Germany. He has held senior positions with BHP Billiton, National Australia Bank, and BlueScope Steel. He was a Senior Adviser to the Premier of Victoria and worked as a Senior Journalist at the Seven and Nine networks in Australia. He is a Research Fellow at Monash Business School where he is completing a Ph.D. in Environmental Governance, a member of the International Advisory Committee of the Global Water Institute at the University of New South Wales, President of the Strathbogie Ranges Conservation Management Network, and runs a 250ha property for conservation and production. He is an Honorary Life Member of both FSC Australia and AWS Asia-Pacific.

Sangwon Suh is a Professor at the Bren School of Environmental Science and Management at the University of California, Santa Barbara. He earned his Ph.D. in Industrial Ecology at Leiden University in the Netherlands. His research focuses on the sustainability of the human-nature complexity through understanding materials and energy exchanges between them. His work contributed to the theoretical foundations and practical applications of life cycle assessment (LCA) and industrial ecology. Dr. Suh was appointed as a member of the International Resource Panel (IRP) by the United Nations Environmental Programme (UNEP) and served the Intergovernmental Panel on Climate Change (IPCC) as a Coordinating Lead Author. He received the

McKnight Land-Grant Professorship from the University of Minnesota's Board of Regents, Leontief Memorial Prize and the Richard Stone Prize from the International Input-Output Association (IIOA), and the Robert A. Laudise Medal from the International Society for Industrial Ecology (ISIE).

Xiaopu Sun is Senior China Counsel at the Institute for Governance & Sustainable Development (IGSD), in Washington DC, USA, and Paris, France. Sun has over a decade of experience working in the Chinese, international, environmental, and energy-efficiency law areas. At IGSD, Sun works with scientists, economists, lawyers, and other colleagues to bring about global fast-action to address the climate emergency, including the near-term climate mitigation through reducing the emissions of the short-lived climate pollutants (black carbon, tropospheric ozone, methane, and hydrofluorocarbons). She was involved in various projects, including supporting the development of the Kigali Amendment to the Montreal Protocol on Substances that Deplete the Ozone Layer, and strengthening the air pollution law enforcement in China. Sun holds an LL.M. from Harvard Law School and an LL.B. in Law and an LL.M. with a concentration in Environmental Law from Peking University Law School in Beijing.

Yixian Sun is a Lecturer (Assistant Professor) in International Development at the University of Bath, UK. His research areas include transnational governance, environmental politics, and sustainable consumption with a focus on the Global South. His work seeks to explain the potential of different governance tools and processes to help emerging economies achieve sustainability transition. Yixian has published peer-review articles in high-impact journals such as *Ecological Economics*, *Global Environmental Politics*, *Global Food Security*, *Nature Food*, and *Review of International Political Economy*. He is finishing a monograph on transnational eco-certification in China. A Chinese national, Yixian holds a Ph.D. and M.A. in International Relations/Political Science, from the Graduate Institute of International and Development Studies in Geneva and B.A. from Nanjing University, and was Postdoctoral Fellow at Yale University.

Bowen Tan is Project Manager for the Paradise Foundation Blue Initiative Fund, the first grant-making fund dedicated to marine conservation in China. In this capacity, Bowen facilitate the Fund to support China's budding marine conservation organizations, assists leadership in shaping the strategic direction of the Fund, and oversees its day-to-day operations. Prior to the Paradise International Foundation, Bowen served as Research Director for the Beijing LGBT Center and as Green Consumption Project Supervisor at the Institute of Public and Environmental Affairs. Bowen received the 2019 Richard Rockefeller Fellowship for his work. He holds a Master of Arts in Sociology from the University of Chicago and a Bachelor of Laws in Sociology from Renmin University of China.

Mr. Xin Wang is the Vice President and a partner of SynTao Co., Ltd. He has extensive experience in helping companies in China with sustainability-related services such as CSR strategy and planning, CSR reporting and training, ESG management and disclosure, stakeholder engagement, CSR program designing, monitoring and evaluation, and community investment.

Before joining SynTao, Mr. Wang worked for the Calvert Foundation in Washington DC, a socially responsible investment institution focusing on supporting grassroots NGOs with innovative financial vehicles. He had also volunteered for the Grameen Foundation, USA, to help with the promotion of microfinance in China to alleviate poverty.

Mr. Wang obtained a Master of Pacific and International Affairs degree from University of California, San Diego.

Zhenzhen Xu is Regional Manager for the Alliance for Water Stewardship Asia-Pacific (also known as Water Stewardship Australia) based in Shanghai. She previously worked as co-leader of the China Water Program for the International Finance Corporate (IFC), part of the World Bank Group; as a business development officer and project manager with Veolia Water in China; and as a water engineer with Sogreah Environmental Protection Consultancy. She holds a Bachelor of Environmental Engineering from Tongil University (Shanghai) and a Master of Environmental Management and Development from the ANU in Canberra.

Oran R. Young is Professor emeritus at the Bren School of Environmental Science and Management at the University of California, Santa Barbara. He is one of the leaders in contemporary thinking about governance and has written extensively about the role of social institutions as core elements of governance systems operating in both international and domestic settings. He is well known both for his theoretical work on governing complex systems and for his applied work dealing with issues of governance relating to the polar regions, the oceans, and the atmosphere. In recent years, he has devoted increased attention to governance systems operating at the domestic level with an emphasis on comparing and contrasting the institutionalized governance processes that have developed in China and the USA.

Rodrigo Zeidan is Associate Professor of Practice of Business and Finance at New York University Shanghai and Visiting Professor at Fundação Dom Cabral. Professor Zeidan is the author of *Economics of Global Business* (MIT Press), and Tariff reduction on renewables inputs for European decarbonization in Nature Sustainability. His research has also been published in some top journals in finance and economics, such as *Journal of Corporate Finance*, *Energy Economics*, *Harvard Business Review*, *International Journal of Production Economics*, and *Journal of Business Ethics*. Rodrigo has a biweekly column at Folha de S. Paulo, the largest Brazilian newspaper, and he has written extensively for international media outlets, including the New York New Times, Bloomberg, and Americas Quarterly. Companies he works or has worked with include Kraft Foods, Johnson & Johnson, Santander, Skanska, L'Oreal, RCI Banque, Vale, Itaú Bank, and Petrobras. His website is: http://rze idan.com.

Juan Zhang has joined the Office for Social Responsibility of China National Textile and Apparel Council (CNTAC) since 2014 and has been mainly responsible for Voluntary sustainability standards research and projects management on the topics of corporate social responsibility and responsible supply chain.

During her working journey, Juan Zhang provides an abundant of training on transparency and capacity building to companies in China textile and apparel sector. Juan Zhang also drafts policy recommendations regarding labor rights, responsible supply chain to China governments.

Juan Zhang is invited as expert consultant by China Chamber of Commerce of Metals Minerals & Chemicals Importers & Exporters (CCCMC) to join the standard-setting process of *Chinese Due Diligence Guidelines for Responsible Mineral Supply Chains* and *Guidance for Sustainable Natural Rubber*. Juan ZHANG is also project expert to China Association for Standardization (CAS) and National Small & Medium Enterprises Council (NSMEC).

List of Figures

The Landscape of Non-state Actors and China Environmental Governance: Illustrative Roadmaps to Processes and Institutions

The Governance Effect of Environmental CSR Reporting in China: State and Non-State Facilitation

Will China Industrial Organizations Succeed in Addressing the "Trust Deficit" Arising from Global Supply Chain Governance?

Extending Enforcement: How the Institute of Public and Environmental Affairs Leverages Public Information to Strengthen Environmental Governance

Water Stewardship: Engaging Business, Civil Society and Government in Collaborative Solutions to China's Freshwater Challenges

In the Shadow of the State: The Rise and Limits of Transnational Private Certification in China's Seafood Sector

Green Supply Chain Initiatives in China: The Roles of Nonstate Actors

List of Tables

Extending Enforcement: How the Institute of Public and Environmental Affairs Leverages Public Information to Strengthen Environmental Governance

In the Shadow of the State: The Rise and Limits of Transnational Private Certification in China's Seafood Sector

Green Supply Chain Initiatives in China: The Roles of Nonstate Actors

An Introductory Framework for Researchers and Practitioners

Introduction

Dan Guttman, Yijia Jing, and Oran R. Young

The Central Theme

This book grew out of a December 2016 workshop at Fudan University in Shanghai. The workshop brought together a mix of practitioners, scholars, and "pracademics" (practitioner/scholars) from China, the United States, the UK, Australia, and Brazil with, collectively, many

D. Guttman (✉)
Tianjin University Law School, Tianjin, China
e-mail: dg109@nyu.edu

Fudan University Institute for Global Public Policy, Shanghai, China

New York University US Asia Law Institute, New York, USA

Y. Jing
Institute for Global Public Policy, School of International Relations and Public Affairs, Fudan University, Shanghai, China
e-mail: jingyj@fudan.edu.cn

D. Guttman et al. (eds.), *Non-state Actors in China and Global Environmental Governance*, Governing China in the 21st Century,
https://doi.org/10.1007/978-981-33-6594-0_1

years of work on environmental affairs in China and global environmental governance. The objective was to explore the relevance of the western concept of nonstate actors to understanding China's experience both in domestic environmental governance and in transnational and even global environmental governance.

In the late twentieth century, the concept of "nonstate actors" gained currency in western thinking about governance, both at the domestic level and in regard to the development of international regimes. The premise was that we cannot rely exclusively on governments to meet a variety of needs for governance, including matters of environmental protection. Both practitioners and scholars began to focus on the roles that nonstate actors play either in stimulating and monitoring the actions of governments or, in some cases, addressing needs for governance in the absence of action on the part of governments.

Attention focused on two distinct types of nonstate actors: (i) nongovernmental organizations often referred to as NGOs or NPOs (nonprofit organization), such as the (US based) Natural Resources Defense Council and the Worldwide Fund for Nature and (ii) profitmaking corporations and business-oriented organizations, such as Unilever and the International Organization for Standardization. In some cases, the two types of nonstate actors have joined forces. In the United States, for example, the Environmental Defense Fund has formed partnerships with major corporations to address environmental issues. The Marine Stewardship Council, an independent nonprofit organization, grew out of a partnership between the Worldwide Fund for Nature and Unilever. The distinction between state and nonstate actors also differs from one society to another and may not be entirely clear in specific cases. In the United States, environmental NGOs play an active role in seeking to influence legislative processes; they also intervene in

O. R. Young
Fudan University, Shanghai, China
e-mail: young@bren.ucsb.edu

Bren School of Environmental Science and Management, University of California Santa Barbara, Santa Barbara, CA, USA litigation making its way through federal and state court systems. In other systems, the legal status of nonstate actors is unclear. For example, they may be established formally as private corporations or effectively directly by the state: an examination of their actual practices often reveals complex and ambiguous connections to government agencies.

Over the last 25 years, a sizable body of literature analyzing the role of nonstate actors in a variety of issue areas has arisen. The role of these actors in the realm of environmental affairs has drawn the attention of a number of prominent scholars. Analysts have sought to explain the origins of the growing roles of nonstate actors. They have identified specific mechanisms that these actors employ in seeking to influence behavior in a variety of situations. Above all, they have asked whether the intervention of nonstate actors has made a difference in efforts to solve environmental problems.

For the most part, this literature is western in origin; it is associated with the concept of governance (the notion that solutions to public problems require the energy of nonstate actors as well as government) and with the analysis of regimes, a concept that took root among western scholars during the last several decades of the twentieth century and that has flourished since (Krasner 1983). Most empirical applications have addressed initiatives unfolding in the United States and Europe or focused on western-driven international regimes dealing with matters like the depletion of the stratospheric ozone layer or the conservation of species and ecosystems. While the resultant literature has devoted some attention to problems arising in other parts of the world (e.g., the destruction of the Amazon rainforest, threats to African elephants), work in this field has remained closely tied to western legal practices and western approaches to research. There have been few applications dealing with issues arising in China or endeavoring to map western thinking about the roles of nonstate actors onto Chinese practices and perspectives pertaining to governance.

All now recognize this as a major gap. Spectacular economic growth over a period of four decades coupled with globalization has turned China into a critical element in global supply chains in a wide range of fields. China is now the largest emitter of greenhouse gases by a wide margin. The development of China's Belt and Road Initiative is turning

China into a major player in economic development ranging from Africa to the Arctic. Efforts to address a wide range of environmental problems cannot succeed without engaging China effectively. To do so, we must gain a better understanding regarding Chinese thinking about governance in general and about the nature and roles of nonstate actors in particular.

At the time of the 2016 workshop, some were already thinking and writing about the roles of global and domestic NGOs in advocacy campaigns in China, and also about what are called "environmental NGOs" in China. But we knew little about the roles of public service units (shiye danwei), a type of institution that is prominent in China but has no clear-cut western analog, or about the growing sustainability focus of business or industrial associations (a form of shehui tuanti or social group or organization), an increasingly important type of actor that operates in close cooperation with government agencies in China. In effect, any effort to understand the role of state and nonstate actors in environmental governance in China would need to come to grips with the nature of the institutionalized governance processes operating in China and the implications of differences between these processes and their western counterparts for our understanding of the roles of nonstate actors (Young et al. 2015).

How, participants in the 2016 workshop asked, were global nonstate actors working in China interacting with Chinese enterprises and government agencies? Given the core role of the state and party in China today, what has emerged to occupy the role or space (kong jian) occupied in western environmental governance by nonstate actors? What are key institutional forms that comprise the nonstate actor landscape, and how do they differ from western models? Now, as Chinese enterprises and investment go abroad, how will rules or guidelines being set by Chinese nonstate actors affect global environmental governance? Is a new China model developing? If so, what can we learn from it?

In the remarkable post-1978 reform and opening up period, Chinese officials and scholars studied and learned from the West, often as students in western universities. Following the end of the cold war, there was a widespread view among western scholars (and leaders) that as China developed markets and reentered the global system, Chinese and western governance systems would gradually converge, albeit with "Chinese characteristics" applied to western models. In Beijing and Shanghai universities and (English language) public policy conferences and workshops, "governance" themes, such as "rule of law," "markets," and "civil society organizations," became common topics. By 2012, the China Daily (a government-sponsored English-language paper) reported that China had nearly 500,000 "NGOs" (*China Daily* 2012). The term "governance" itself is now a core Chinese expression; the English title of Xi Jinping's multi-volume work is "The Governance of China" (Xi 2014, 2017, 2020).

By the time of the 2016 workshop, however, it was becoming apparent that, although Chinese scholars and officials teach and use the same core terminology to talk about governance (often expressed in English), western and Chinese governance systems are not converging as many western scholars and leaders had anticipated (Fukuyama 1992). Thus, it is imperative to understand whether and how well western terms like nonstate actors (or "NGO" or "CSR") translate into the governance processes of the twenty-first-century Chinese "operating system." In a world where challenges transcend borders, the need for comparative understanding, and shared learning, is greater than ever (Zhao et al. 2020).

From this perspective, several general observations have emerged from the work growing out of the Fudan Workshop and reported in the chapters included in this book. There is no sharp separation between state and nonstate actors in China. All those working on environmental issues in China operate in the "shadow of the state." The Institute of Public and Environmental Affairs (IPE), often regarded as a western-style NGO, uses data generated by government agencies and engages in complex public–private interactions. Other important actors, including "shiye danwei" and a number of types of business associations, have no direct western counterparts. They have close links to the state, but they can and do play significant roles in nudging both private and state-owned enterprises to adopt and adhere to environmental standards. Given the distinctive features of the Chinese "operating system," it is also important to emphasize the differences between western and Chinese practices regarding the mechanisms through which those concerned with environmental protection endeavor to achieve influence. Whereas U.S. nonstate actors tend to make use of adversarial processes, Chinese actors typically participate in planning processes, provide resources to help enterprises meet goals articulated in plans, and assist in the development of new practices in the aftermath of crises. Comparing the two systems in terms of performance is a challenging task that we are just beginning to tackle. But it is clear already that we cannot say that one system is better than the other in some general sense when it comes to addressing environmental problems.

Since 2016, the building blocks of the late twentieth-century western-inspired global order are daily being called into question. Depending on one's perspective, core developments include the Trump administration's declared withdrawal from the Paris Climate Agreement, the Biden Administration's vigorous reaffirmation of US commitment to addressing climate change with other countries, and broader challenges to the global trading system and China's assumption of a growing global role through, for example, the Belt and Road Initiative and the Asia Infrastructure Investment Bank. By 2020, talk of the ending of the late twentieth-century global order and the "decoupling" of US-China relations was common.

Now there is COVID-19. The pandemic punctuates most dramatically the need for shared understanding of major governance systems, especially where there are differences among them but cooperation and shared learning are essential. What is needed is a commitment to rigorous comparative analysis both to gain insight into the determinants of success in different governance systems and to devise new approaches to cooperation where solving environmental problems requires effective cooperation between and among societies whose governance processes differ in important ways. We hope the chapters in the book provide a framework for those who continue to work on such matters and will be working in years to come to repair and or restructure global environmental governance systems.

The Structure of the Book

The body of this book comprises three interrelated parts. Part I (chapters "Introduction" to "The Landscape of Nonstate Actors and China Environmental Governance: Illustrative Roadmaps to Processes and Institutions") provides a framework for understanding and conducting further research on China's emerging system of nonstate actors in environmental governance. The chapters in Part II (chapters "The Governance Effect of Environmental CSR Reporting in China: State and Non-state Facilitation" to "Extending Enforcement: How the Institute of Public and Environmental Affairs Leverages Public Information to Strengthen Environmental Governance") focus on key types of Chinese environmental nonstate actors—profit-making enterprises, business associations (*she hui tuan ti or hang ye xie hui*), and environmental "NGOs." Part III (chapters "Water Stewardship: Engaging Business, Civil Society and Government in Collaborative Solutions to China's Freshwater Challenges" to "Green Supply Chain Initiatives in China: The Roles of Nonstate Actors") contains case studies of the ways in which Chinese and global nonstate actors have interacted and continue to interact in the development of environmental governance systems. Broadly speaking, as China became the world's factory, global nonstate actors drove changes within China. Now, as China's own global development initiatives have increased, native Chinese nonstate actors are beginning to develop standards both for domestic use and for Chinese enterprises that go abroad.

Following this introduction, chapter "The State, Nonstate Actors, and China's Environmental Performance: Setting the Stage", co-authored by participants in the 2016 Fudan Workshop, provides a framework for thinking about China nonstate actors engaged in environmental governance in the context of and by comparison with the global landscape of nonstate actors. The chapter begins with a review of the Chinese environmental nonstate actor "institutional landscape." This landscape now includes "global" nonstate actors at work in China, and Chinese institutions such as business enterprises,[1] public service units (*shiye dan wei*), and social groups (*she hui tuan ti*), including business or trade associations. It also encompasses nonprofit enterprises (*minban fei qiye*) and foundations (*jijin hui*) that are sometimes analogized to western

NGOs (or NPOs). The chapter then asks: how do institutions that are not government agencies seek to have effect in making rules work? To address this question, the chapter introduces two concepts to help understand nonstate actor strategies and choices: streams of activities and mechanisms of influence. The chapter differentiates several "streams of activities" including (1) institutions aiding governments in enforcement of government set rules; (2) "nonprofit" non-state actors engaging in rule or guidance setting and enforcement; and (3) profit-making enterprises engaging in rule or guidance setting and enforcement on their own. Regarding mechanisms of influence, the chapter identifies five types of levers or strategies nonstate actors use to exercise influence over the environmental and sustainability relevant conduct of enterprises. The chapter asks what can we say at this relatively early stage about the effectiveness of nonstate actors in attaining their goals and concludes with insights relating to the further evolution of Chinese nonstate actors in environmental governance domestically and as China goes abroad.

Chapter "The Landscape of Nonstate Actors and China Environmental Governance: Illustrative Roadmaps to Processes and Institutions" portrays institutional landscapes by providing multiple views of differing landscapes from a visual as well as narrative perspective. The chapter, whose primary author is Ruoyu Chen, a recent New York University graduate soon embarking on graduate study at Yale, complements subsequent chapters with graphic portrayals of the landscapes of actors engaged in standard making, certification and supervision and water, air, textile, and (climate change related) air conditioning regulation.

Moving to Part II, chapter "The Governance Effect of Environmental CSR Reporting in China: State and Non-state Facilitation" by Lydia Price, a professor and director of sustainability programs at the China-European International Business School (CEIBS) in Shanghai, reviews the corporate social responsibility (CSR) rules and performance of Chinese enterprises. CSR, a term that originated in the West, is now a global concept. In its late twentieth-century western origins, the concept involved voluntary efforts by business and was sometimes thought of as business self-regulation. Over time, organizational CSR practices and programs have been shaped by various external forces, including market demands, community pressures, and government guidelines. As Price explains, in its twenty-first-century embrace by China, the concept is strongly shaped by government direction of corporate activity. In China, government rules and policies directly call for CSR from state-owned and publicly listed companies. Further, global and domestic non-state actors play important roles in the implementation of Chinese CSR rules. These include Chinese shiye danwei and foreign/global non-state actors with a CSR focus. The chapter draws on Price's rich compilation of data and analysis of Chinese environmental CSR reporting to date to provide perspective on whether and how practices of environmental CSR are emerging.

In chapter "Will China Industrial Organizations Succeed in Addressing the "Trust Deficit" Arising from Global Supply Chain Governance?", Juan Zhang draws on her years of research and practical experience with the China National Textiles and Apparel Council (CNTAC) and other China business associations to focus on the evolving roles of Chinese business or trade associations (primarily shehui tuanti) in the development of sustainability standards. In the late twentieth century, as China became the world's manufacturing center, global non-state actors and multinationals introduced "VSS" (voluntary sustainability standards) into China through Chinese legs or components of global supply chains. In the twenty-first century, China shehui tuanti are themselves beginning to set standards. Moreover, as Chinese enterprises are going abroad, the guidelines or standards Chinese shehui tua ti are setting are now being applied to Chinese activities (e.g., such as mining) abroad. With the understanding that the role of *shehui tuanti* in sustainability is a work in progress, Zhang outlines the evolving institutional landscape, challenges, and the strategies, dynamics, and challenges that may lead to success in the efforts of Chinese *shehui tuanti* to develop voluntary standards.

In chapter "Extending Enforcement: How the Institute of Public and Environmental Affairs Leverages Public Information to Strengthen Environmental Governance", Kate Logan focuses on the work of the Institute for Public and Environmental Affairs (IPE), a *minban fei qiye* often thought of as a western-style NGO. Logan's work draws on her experience both as a member of the IPE staff and as a graduate of Yale's School of the Environment. IPE was founded by Ma Jun, renowned globally as one of China's leading environmental activists. During study at Yale as a Yale World Fellow, Ma Jun learned of the role of "information-based regulation" in environmental governance. On his return to China, he founded IPE, which has taken a lead role in promoting the use of environmental information as a means to assure compliance with laws and policies. As Logan's chapter explains, IPE plays a unique middleman role in China's landscape of environmental governance by leveraging government-published information to improve government and corporate environmental compliance alike. IPE relies on existing government policies and information to develop the standards and indices that it implements. These indices supplement government environmental enforcement efforts where capacity may be lacking, namely, to supervise local-level government environmental disclosure as well as the environmental compliance of corporations. The chapter reviews the development of IPE's approach to public (web-based) databases and reports that have become standard resources in China (and globally for those interested in China), and considers the institutional mechanism(s) of influence that underlie the development.

Moving on to Part III, chapter "Water Stewardship: Engaging Business, Civil Society and Government in Collaborative Solutions to China's Freshwater Challenges" by Michael Spencer and Zhen zhen Xu reports on and analyzes the impact of efforts of the Alliance for Water Stewardship (AWS) to work with China local governments and water industry actors in the introduction of sustainable procedures and techniques. The chapter draws on case studies from their work as Officer and Asia director of AWS, including Spencer's work in Australian business and research at Monash University and Xu's work in consulting and banking and further work throughout the Asia Pacific region. AWS is an international membership-based civil society organization, with multi-stakeholder governance, offering a global standard for water stewardship, a verification system, and a brand that allows water-users to communicate responsible management of water resources. The AWS Standard, developed by a multi-stakeholder International Standard Development Committee, is designed to address societal concerns about water: scarcity, pollution, ecosystem health, cultural use of water, equity of access to water, sanitation, and health services related to water. Spencer and Xu draw on pilot projects using the AWS Standard with local government and industry in a Tianjin (northern China) economic development zone and in Kunshan City (in the Yangtze River Delta's Jiangsu Province), to develop and test hypotheses about the role of nonstate actors, including mechanisms of influence, in gaining Chinese enterprise acceptance of sustainability standards.

In chapter "Non-industry and Nonstate Actors Contribution in the Standard Drafting Process: Examples from the Development of China Room Air Conditioner Standards", Richard ("Tad") Ferris and Xiaopu Sun provide a primer on global non-state actor expert participation in Chinese government rulemaking. The co-authors are China law and environment experts and practitioners. Ferris brings decades of private law practice experience advising enterprises, multilateral institutions, and nonprofit organizations on Chinese and international environmental, health, safety, and climate law, including product standards and supplier compliance. Sun received law degrees from Peking University and Harvard University and has over a decade of experience working in China, international, environmental, and energy-efficiency law areas. Ferris and Sun, now at the Institute for Governance & Sustainable Development (IGSD), work with scientists, former government officials, and other colleagues to bring about global fast-action to address the climate emergency. IGSD helped achieve adoption of the 2016 Kigali Amendment to the Montreal Protocol on the phasedown of HFCs (hydrofluorocarbons), which can avoid up to 0.5 °C of warming by the end of this century. Improvements to the energy efficiency of cooling equipment could perhaps double this environmentally necessary achievement. The HFCs phasedown and cooling efficiency improvement require among other steps, review and revision of domestic air conditioner and refrigerant policies and standards. Drawing on their substantial practical and scholarly experience, the chapter provides perspectives on the role of global nonstate actor experts in the Chinese government's energy-efficiency standard-setting processes.

In chapter "In the Shadow of the State: The Rise and Limits of Transnational Private Certification in China's Seafood Sector", Yixian Sun, a Chinese scholar based at the University of Bath in the UK, focuses on the rise and limits of transnational private certification procedures in China's seafood industry. Sun draws on his years of research, including field interviews. This chapter chooses seafood as a critical case to understand the rise of transnational certification schemes in China's unique governance processes dominated by the state. The analysis shows pathways of the rise of transnational certification schemes and interaction among relevant stakeholders in China's seafood sector. Rather than relying on activist campaigns as proponents have done in other settings, private governance organizations and their supporters have engaged domestic industry associations actively in China to enhance the legitimacy of their rules and generate interest among downstream businesses in the Chinese market. While this strategy has raised the awareness of Chinese stakeholders about the concept of sustainability and produced some incremental progress, private governance alone seems unlikely to move the whole seafood sector toward sustainability without supportive public policies. Better integration between public and private governance will be needed to promote a sustainability transition in China's seafood sector.

Finally, in chapter "Green Supply Chain Initiatives in China: The Roles of Nonstate Actors", Lingxuan Liu, a Chinese scholar now based in the UK at the University of Lancaster, focuses on Chinese government's "nonregulatory" efforts to address multinational supply chains. Liu draws on his research and also on his work as advisor to multinational companies on green supply chain management. With a focus on China's program of Green Supply Chain Initiatives (GSCIs) introduced in 2014, the chapter examines four pioneering cases involving Chinese cities and a cohort of manufacturing firms that have become the pioneers of specific GSCIs. The chapter begins with an analytical framework to explore the design and implementation of the initiatives. This exploration aims to generate new understandings of the roles of nonstate actors in environmental governance as well as the specific mechanisms nonstate actors use to influence environmental practices in both the public sector and the private sector.

Note

1. We recognize that many of China's, and the world's, largest profit-making enterprises are Chinese state-owned enterprises (*guoyou qiye*), thus not "non-state actors." These include oil, electric energy, mining, chemical, construction, and other companies that are core foci of environmental concern. Chapter authors note state-owned status where relevant to discussion.

References

China Daily. 2012. Number of NGOs in China Grows to Nearly 500,000. March 20. https://www.chinadaily.com.cn/china/2012-03/20/content_14875389.htm.

Fukuyama, Francis. 1992. *The End of History and the Last Man*. New York: The Free Press.

Krasner, Stephen D. (ed.). 1983. *International Regimes*. Ithaca: Cornell University Press.

Xi, Jinping, 2014, 2017, 2020. *The Governance of China*, 3 vols. Beijing: Foreign Language Press.

Young, Oran R., Dan Guttman, Ye. Qi, Kris Bachus, David Belis, Hongguang Cheng, Alvin Lin, et al. 2015. Institutionalized Governance Processes: Comparing Environmental Problem Solving in China and the United States. *Global Environmental Change* 31: 163–173. https://doi.org/10.1016/j.gloenvcha.2015.01.010.

Zhao, Xiaofan, Oran R. Young, Ye Qi, and Dan Guttman. 2020. Back to the Future: Can Chinese Doubling Down and American Muddling Through Fulfill 21st Century Needs for Environmental Governance? *Environmental Politics and Governance* 30 (2): 59–70. https://onlinelibrary.wiley.com/doi/abs/10.1002/eet.1884.

The State, Nonstate Actors, and China's Environmental Performance: Setting the Stage

Dan Guttman, Yijia Jing, and Oran R. Young

Introduction

In the West, the limited capacity of governments to make and enforce environmental standards has led to an increase in the efforts of a variety of "nonstate actors" to provide alternatives, or at least supplements, to governmental actions as a means of addressing environmental problems.

This chapter is a version of an article originally published in the *Journal of Environmental Management*, which we thank for permission to reprint. The original article appears as Dan Guttman, Oran Young, Jing Yijia, Barbara Bramble, Maoliang Bu, Carmen Chen, Kathinka Furst, Tao Hu, Yifei Li, Kate Logan, Lingxuan Liu, Lydia Price, Michael Spencer, Sangwon Suh, Xiaopu Sun, Bowen Tan, Harold Wang, Xin Wang, Rodrigo Zeidan, Juan Zhang, Xinxin Zhang, "Environmental Governance in China: Interactions Between the State and "Nonstate Actors"; *Journal of Environmental Management* 220: 126–135 (2018).

D. Guttman (✉)
Tianjin University Law School, Tianjin, China
e-mail: dg109@nyu.edu

Fudan University Institute for Global Public Policy, Shanghai, China

D. Guttman et al. (eds.), *Non-state Actors in China and Global Environmental Governance*, Governing China in the 21st Century,
https://doi.org/10.1007/978-981-33-6594-0_2

This category of actors includes both environmental nongovernmental organizations (ENGOs) and corporations that, individually or in association, take an interest in the systematic modification of enterprise behavior to embrace sustainability. In many prominent cases, the focus is on global supply chains, and the relevant nonstate actors include international groups that reach across state borders.

In this chapter, we ask how the rapidly developing western literature on the roles that nonstate actors play in addressing environmental problems applies to the experience of China and provide a framework for responding to this question. Are the key findings and frameworks associated with this literature generalizable from western systems to the distinctive setting that has evolved in China during the nearly four

In addition to Guttman, Jing, and Young, the further authors of this chapter are: Barbara Bramble (National Wildlife Federation); Maoliang Bu (Nanjing University School of Business); Kathinka Furst (Duke Kunshan University); Tao Hu (WWF, USA); Yife Li (Shanghai New York University); Sanwun Suh (University of California Santa Barbara Bren School of Environmental Science and Management); Bowen Tan (Institute of Public and Environmental Affairs, Beijing, China); Xin Wang (Syntao; Beijing, China); Xinxin Zhang (Rainforest Alliance, Beijing, China); Rodrigo Zeidan (Shanghai New York University; Fundacao Dom Cabral, Brazil). This chapter is the product of a workshop convened to include presentations and discussion among university-based scholars and global and China "nonstate actor" practitioners. As such, the co-authors, as noted, include individuals affiliated with some of the organizations discussed in the text. The three university-based lead authors (Guttman, Young, Jing) do not have affiliations with any of the "nonstate actor" organizations addressed in the article.

D. Guttman
New York University US Asia Law Institute, New York, USA

Y. Jing
Institute for Global Public Policy, School of International Relations and Public Affairs, Fudan University, Shanghai, China
e-mail: jingyj@fudan.edu.cn

O. R. Young
Fudan University, Shanghai, China
e-mail: young@bren.ucsb.edu

Bren School of Environmental Science and Management, University of California Santa Barbara, Santa Barbara, CA, USA decades since the reform and opening up of the Peoples Republic of China (PRC) began? Do we need to make significant adjustments in the analyses articulated in this literature to explain developments in China? Equally important, we also ask whether we can derive lessons from the Chinese experience that will broaden and deepen our general understanding of the roles nonstate actors play in environmental protection. We aim to advance our fundamental understanding of environmental governance and at the same time to contribute to the thinking of those engaged in applied efforts to address environmental problems in China and elsewhere (Young et al. 2015).

There is a growing literature on both the domestic and the cross-border activities of nonstate actors relating to environmental issues in western democracies and developing countries in Africa, Asia, and South America (Bartley et al. 2015; Auld 2014; Green 2014; Buthe and Mattli 2011; Cashore et al. 2004). However, there is little literature on whether and how global nonstate actor networks operate in or with China, and on the landscape of a China's native counterpart community. This literature encompasses two broad streams, one dealing with the role of environmental NGOs and business associations in creating standards, certification schemes, and codes of conduct and seeking to ensure their implementation, and the other dealing with the role of corporations and focusing on what analysts often characterize as corporate social responsibility or CSR (Baron 2009; Vogel 2006).

For the most part, this western literature does not deal with the experience of China, despite the fact that China accounts for ~18% of the Earth's human population, is now the world's second largest economy, and is at the center of a burgeoning network of global trade relationships. Equally important, China has a distinctive system in which there is no clear separation among the public sector, the private sector, and civil society and central planning remains a key mechanism for setting priorities and allocating society's resources.

Our central argument is that the current western vernacular is not robust enough to capture the realities of what is happening in China where government is the central player in all realms and western distinctions among the public sector, the private sector, and civil society are not clearly applicable. China has its own institutionalized governance processes (Young et al. 2015). Dealing with environmental protection in this setting involves a set of practices that cannot be understood through a lens that features a sharp distinction between the state and nonstate actors. Any effort to shed light on the roles that various types of actors play in addressing issues of environmental protection in China must take the differences between these Chinese institutionalized governance processes and familiar western processes as a point of departure.

To develop this argument, we proceed in several steps. In the first substantive section, we provide context regarding the most important differences between western systems and the realities of contemporary China. We then ask: how does the western concept of nonstate actors translate into the Chinese context (Guttman 2015)? Chinese scholars and practitioners are familiar with, and often use, western terms like "NGO" and "CSR." But these terms take on distinct meanings when used with reference to Chinese governance processes. While we retain the English phrase "nonstate actor" for purposes of analysis in this article, we emphasize that the western usage of the term does not map onto the Chinese experience precisely.

With a clearer picture of the relevant actors in hand, we proceed in the next section to examine the main streams of activities on the part of nonstate actors to participate in environmental governance, with agendas emanating from the state, from organizations independent of enterprises, and from enterprises themselves, either individually or through associations. This sets the stage for an examination of the principal mechanisms through which these actors endeavor to influence the behavior of major players. In the following substantive section, we explore the results of the efforts of nonstate actors to influence the course of environmental governance in China and make a preliminary effort to identify key determinants of success.

In the final substantive section, we assess both limits on the applicability of the western literature on nonstate actors to understanding environmental governance in China and insights from the Chinese experience that may enrich the analysis of the role of nonstate actors in environmental governance in global terms. We conclude with some reflections on next steps in the analysis of the role of nonstate actors in environmental governance.

Setting the Stage: The Landscape of Environmental Governance in China

It is helpful to begin with a brief characterization of the landscape of environmental governance in China. Though China's remarkable economic development features the rise of private enterprises and markets, China is, in the language of the PRC's Constitution, a "socialist" system and in practice one in which the government (including the Communist Party of China) plays the key role in directing the economy and related environmental developments. While China has been marketizing, state planning—from the central Five Year Plan (FYP) to myriad further central and local plans—continues to set core goals for society and the allocation of resources. The concept of ecological civilization (*shengtai wenming*), introduced at the 17th Party Congress in 2007, has been accompanied by what has been called the greening of the Five Year Plan (Koleski 2017; Young et al. 2015; CPC 2016).

Pollution, treated as an "externality" in many western analyses of environmental governance, is internalized in the planning process in the Chinese governance system. Core enterprises, including energy companies that are key sources of pollution and among the world's largest enterprises, are state-owned enterprises or SOEs (*guoyou qiye*). The government also owns much of the land and natural resources at issue in sustainability.

Over the millennia, China's governments have been civilizational leaders in standard setting. Despite its short reign, the Qin dynasty (221–206 BCE) is considered a seminal dynasty in part because of its role in standardization of weights, measures, coins, written script language, and administrative structures (Lewis 2007). In the early PRC era, standard setting was under government direction. In the "opening up and reform period," from the 1980s under Deng Xiao Ping, as enterprises and markets developed, government agencies for standard setting and supervision were created. These include, at the ministerial level, the General Administration of Quality Supervision, Inspection and Quarantine (AQSIQ), and under AQSIQ, the Standards Administration of the Peoples' Republic of China (SAC) and the Certification and Accreditation Administration of the People's Republic of China

(CNCA). Standards in China fall into multiple categories: national standards, industry standards, local or regional standards, group standards, and enterprise standards for individual companies. National standards can be either mandatory (*guojia biaozhun* or "GB") or voluntary (ITA website).

Beginning with Deng Xiao Ping era "opening up and reform," China's party and government system spawned the development of numerous private enterprises and also hundreds of thousands of organizations conventionally referred to by the government (in English) as "NGOs" or "social organizations" (*China Daily* 2012). Their employees are not "civil servants" (i.e., as provided for by the billeting or *bianzhi* system for civil service positions), and the organizations lack the formal status of "government" organizations. However, they retain strong ties, both formal and informal, to the government (and party).

In this context, standard setting in relation to the environment as well as elsewhere has significantly devolved to actors for which there is often no clear western analog in the "nonstate" actor context. As enterprises and universities developed, the Standardization Law of the People's Republic of China provides (Article 12) that "[t]he roles of trade associations, scientific research institutions and academic organizations shall be brought into play in the formulation of standards." As a 2016 paper by experts affiliated with Tsinghua University and the China National Institute of Standardization (both "*shiye danwei*" as we will discuss) puts it: "[A]fter the [1980's] reform and opening up, all levels of government have promoted the establishment of many standardization associations....These associations are generally not really independent social organizations and have a very close relation with the government" (Wang and Zheng 2016).

Ongoing China standard setting reform seeks to continue to draw on government directed "delegation" to such institutions. This includes group or community standards developed by business associations, which are increasing in number and importance (National Group Standards Information Platform 2016). Article 6 of the Draft Amendment to the National Standardization Law provides that "[t]he government encourages enterprises, social groups, and educational and research organizations to conduct or participate in the standardization work" (Draft

Amendment to Standardization Law). In addition, the reform calls for cooperation with, and learning from, "foreign" standard setting institutions, toward the development of standards suitable to China's characteristics (Circular of the State Council 2015). In a related vein of calling on "NGOs," in 2014 China amended its environmental law to provide that environmental NGOs (that meet criteria of longevity, charter mission, and good standing) can represent the "public interest" in court actions (UNDP 2015).

For our purposes, the essential point is that in China the government plays a key role in all sectors of society, a fact that makes it misleading to draw sharp distinctions among the public sector, the private sector, and civil society. Though we employ the English phrase nonstate actor in the analysis to follow, it is essential to bear in mind that in China these actors are not entirely autonomous in the sense of being free to adopt strategies without reference to the preferences of the government.

Identifying the Players: The Evolving Community of Environmental "Nontate Actors" in China

The landscape of organizations in China that occupy the territory accorded to nonstate actors endeavoring to promote environmental protection in other societies is populated by a number of types of players.

As a prelude, we note that while terms like "NGO" and "nonstate actor" are widely used in the West, their practical (and legal) meanings may vary from country to country. The bright line they suggest between "state" and "nonstate" is often blurred or a matter of degree. To a Chinese audience, for example, "nongovernmental" may seem hard to apply to a U.S. "NGO" that gains special U.S. tax status [501(c)(3)] on condition of limiting its political activity or to an "NGO" that depends on "government" for much or even all of its funding. Similarly, since the mid-twentieth century, much of the basic work of the U.S. government is performed through grants and contracts to "private" nonprofit and for-profit enterprises. A Chinese scholar might ask, "in what sense is a company like Lockheed, whose year in year out billions of dollars of income are derived nearly entirely from 'government' a "private enterprise"? (Guttman 2000).

"NGOs"—Domestic and Foreign

China has many entities that in the English global vernacular are called "NGOs" (e.g., as referred to in the English language *China Daily* [2012]). These include domestic organizations created by Chinese citizens and China operations of "foreign NGOs."

Domestic NGOs

According to the Ministry of Civil Affairs (MCA), at year-end 2019, there were 866,000 registered social organizations in China (MCA website). Legally registered social organizations include: (1) *minban feiqiye danwei* ("nonprofit enterprises"); "*she hui fu wu ji gou*" ("social service organizations" under the new Charity Law); (2) "*shehui tuanti*" (social groups including, as relevant here, business or professional associations); and (3) "*jijinhui*" (foundations).

In China, the popular appellation "NGOs" is often used for these social organizations, although their relation to government is far from independent or autonomous. In China, the social organization registration and supervision system (SORSS) requires, as a precondition for registration, that a social organization must have a business supervisory agency (*yewu zhuguan danwei*), usually a party or governmental agency (Simon 2013; Jing 2015; Fürst 2016).

Of the hundreds of thousands of registered "NGOs," about 7000 are said to be "environmental NGOs." The vast majority of these are small, with limited staff and resources. Only a few, such as Friends of Nature and the All China Environmental Federation, have national reach. The Institute of Public and Environmental Affairs (IPE), a *minban feiqiye* established by Ma Jun in 2006, is a leading example of a registered "NGO" focused on promoting environmental standards (IPE website). In 2014, China amended its environmental law to provide (at Article

58), as elaborated by a Supreme Court interpretation, that qualified environmental NGOs (e.g., by charter mission, years of operation, and good standing) can represent the public interest in court (Environmental Protection Law; Supreme Court Judicial Interpretation).

"Foreign" NGOs

Since WWF's entry into China in the early 1980s, many ENGOs chartered abroad have come to China. In addition to WWF, as relevant to our story of nonstate actor environmental protection efforts, these include the Forest Stewardship Council, ISEAL Alliance, the Rainforest Alliance, Alliance for Water Stewardship, Marine Stewardship Council, Better Cotton Initiative, Roundtable for Sustainable Palm Oil, and Programme for the Endorsement of Forest Certification (PEFC). They often self-identify as "multistakeholder" organizations, with members from business, "NGOs," and sometimes individuals (WWF 2010). The list of organizations relevant to the story should also include organizations like Greenpeace and Amnesty International, an NGO not present in China, but whose investigatory work, as discussed below, provides entrée for nonstate actor regulatory efforts.

Historically, however, few "foreign" NGOs have been granted legal status under China's NGO law with less than a handful addressing environmental issues. As a consequence, they have sought other status, including registration as business enterprises (*gongsi*) (Simon 2013). Therefore, the 2016 "Law of the People's Republic of China on Administration of Activities of Overseas Nongovernmental Organizations" discussed below is likely fundamentally altering the "foreign NGO's" presence in China. Following the logic of regulations over domestic NGOs, foreign-related NGOs must gain a sponsor from an approved list. In addition, the Public Security Bureau will have core supervisory authority.

Enterprises: CSR

Corporate social responsibility (CSR) is another term in the global vernacular, with particular definitions varying from place to place, even within countries. In one common definition, the concept suggests activities beyond those required by corporate law or government regulatory compliance (Baron 2009; Vogel 2006). In China, following government direction, CSR is now a mainstream Chinese corporate activity, particularly among state-owned and larger publicly traded enterprises with core responsibility for addressing environmental challenges (e.g., China's state-owned oil companies) (CNOOC, Sinopec CSR websites). In export industries, such as apparel and electronics, multinational supply chain codes, as discussed in this article, are reinforcements of CSR.

In 2008, the State Council issued guidelines regarding CSR for state-owned enterprises (SASAC). The Shanghai, Shenzhen, and Hong Kong stock exchanges have issued guidelines and requirements on CSR reporting (including environment) (BSD Consulting, 2016), and the Standardization Administration (SAC) has issued a standard that provides guidelines for voluntary "social responsibility reporting" (Chinese standards net). Of particular note, the "ESG" (Environmental, Social, and Governance) Guidelines issued by the Hong Kong exchange appear to be having a significant impact on smaller Hong Kong-registered mainland enterprises that are not state-owned enterprises governed by SASAC guidelines (Fudan workshop discussion; Hong Kong ESG guidelines).

Business Associations

The PRC's "opening up and market reform" was accompanied by the development of business associations with distinct legal personalities (Hsueh 2011). Typically legally registered as *shehui tuanti* (social groups), but also referred to as *hangye xiehui* (business/industrial associations) or *hangye zuzhi* (business/industrial organizations), these actors are coming to play an increasing role in the promotion of "social responsibility," including environmental standard setting.

Social groups include standard implementers as well as standard setters. The China Certification and Accreditation Association (CCAA) was established in 2005 under AQSIQ (CNCA website). Its work includes implementing national certification and accreditation policies (Article 6) (Id.).

The difficulty of fitting these institutions into the nonstate actor frame is indicated by a 2016 English language presentation by an association's staff, which characterizes them as "public institutions" (Fudan workshop PPT):

> CNTAC [textile industry association] was transformed from the former Ministry of the Textile Industry, and is now a public institution of the State-owned Assets Supervision and Administration Commission of the State Council; CCCMC [mining] is a public institution of the Ministry of Commerce; CESA [electronics] is a public institution of MIIT; the Ministry of Commerce and MIIT are the authorities of CHINCA [contractors] and CACE [communications] respectively.

Shiye danwei (Public Service Units)

Shiye danwei constitute a further organization type relevant to our story that, from the "foreign" perspective, may look like either a "government" or "nongovernment" body but is not properly understood as either. *Shiye danwei* (often translated as public service units) include China's universities, hospitals, research institutes (e.g., the China Academies of Science and Social Science), and organizations, like the China Academy of Environmental Planning, that serve as staff for government agencies. *Shiye danwei* are not government in the sense that staff members are not "civil servants" (*gongwuyuan*). However, given the role of government in their direction (e.g., sponsorship, selection of leadership, assignments), they are not readily cast as "NGOs" (or "nonstate actors") (Guttman et al. 2013; Tang and Lo 2009). In recent years, driven by marketization reform and social reforms, *shiye danwei* have been experiencing a general reform or reclassification. On review of their functions, the status of some has been/is being changed to governmental agency, enterprise, or social organization (Id.).

The China National Institutes of Standardization, for example, is a *shiye danwei* research institution under the AQSIQ. Its English language website describes it as a "public interest institution at national level" (CNIS website). The Chinese Academy of Sciences has issued CSR reporting guidelines (Global Reporting Initiative 2014), and the Chinese Academy of Forestry (a *shiye danwei*) plays a continuing role in developing China's forest certification standards. The China National Accreditation Service for Conformity Assessment (CNAS) (a *shiye danwei*) is responsible for the accreditation of certification bodies, laboratories, and inspection bodies, established under the Certification and Accreditation Administration of the People's Republic of China (CNCA) (CNAS website).

Emerging E-platforms

In their explosive growth, China's three major web platforms—Ali Baba, Tencent, and Baidu—have become the virtual storehouse for services and products both within China and globally. A core question is whether and how these platforms may emerge, as sources for environmental standards for goods and services. While it is premature to begin to answer the question here, the potential role for consumer review and/or product standards provided on websites or by the web platforms is enormous.

Making a Difference: Streams of Activities and Mechanisms of Influence

Observers have chronicled a wide range of strategies that nonstate actors pursuing environmental protection have developed, including common themes and variations among them. To understand the roles such actors play in China, we use the concepts of "streams of activities" and "mechanisms of influence."

With regard to streams of activities, we ask whether the source of the stream is: (1) government rule setting with nonstate actors aiding in implementation; (2) organizations operating outside the government and independent of enterprises, or (3) enterprises acting either individually or through associations to set and implement standards going beyond those established by the state. Are nonstate actors seeking to promote or strengthen the efforts of governments or are they looking for ways to bring about change when governments fail to act?

By mechanisms of influence, we refer to tools or levers nonstate actors employ to alter conduct and the theories of change they reflect. In brief, why should enterprises pay attention to regulators who are not government and lack the penal power of government? And, if they do pay attention, why is it likely to be more than for public relations purposes ("greenwashing")?

Streams of Activities and Institutional Arrangements that Illustrate Their Workings in China

Stream One: Governments or intergovernmental organizations adopt regulatory measures and related standards; nonstate actors play prominent roles in administering or implementing them, ensuring compliance on the part of subjects, or monitoring progress toward achieving goals.

Here, we note examples involving Chinese NGOs and *shiye danwei*.

The Institute of Public and Environmental Affairs (IPE), founded by Ma Jun, a globally recognized China environmental pioneer, is focused on the development of information-based regulation, drawing on government-disclosed required information and developing its own indices and rankings to stimulate compliance with disclosure requirements.

Ma Jun's approach was shaped by time spent in the U.S. and the study of American ENGOs' use of government-mandated environmental information, such as the Toxic Release Inventory data mandated in the wake of the 1984 Bhopal India chemical disaster. In cooperation with the China's program of the Natural Resources Defense Council, IPE developed and operates an annual survey of China's local government environmental information disclosure (the pollution information transparency index or PITI).

In 2011, Ma Jun helped bring global attention to the poisoning of 137 Apple supply chain workers polishing products with N Hexane (Barboza 2011). The PITI index is now accompanied by a corporate information transparency index (CITI), which ranks multinational brands on information disclosure, measuring the "greening of the supply chain" (IPE website). IPE is developing a real estate project to bridge government law enforcement data and realtor readiness to purchase from suppliers who comply with environmental laws.

In addition to IPE, environmental NGOs with broad reach in supplementing governmental enforcement include those authorized by the 2014 amendments to China's environmental law to go to court to represent the public interest. While it is said that hundreds of NGOs may technically qualify (by virtue of environmental mission and five years in good standing), the number of NGOs serving as plaintiffs is small. In 2015, approximately 50 lawsuits were brought, primarily by Beijing-based NGOs including Friends of Nature and the Center for Biodiversity Conservation and Green Development Foundation (Li 2016). Now, the China government procuraturate has been given authority to bring "environmental public interest" lawsuits; it now brings the preponderance of actions (Zhao et al, 2020).

Further, government environmental enforcement is supplemented by an array of accreditation and inspection institutions and *shiye danwei* that serve government in day-to-day monitoring and inspection.

Stream Two: NGOs take the initiative in developing and implementing standards, introducing certification procedures, identifying best practices, or devising codes of conduct, sometimes in partnership with businesses or governments.

Here, the primary actors appear to be foreign nonstate actors engaging in activities in China. These organizations often choose the term "multi-stakeholder" organizations, with reference to members and/or governing boards embracing enterprises, "NGOs," and, even on occasion, citizens. Through training, consultation or other forms of cooperation, they assist China counterparts. In addition, they promote application of their own standards.

The ISEAL Alliance and many of its members exemplify the first approach (ISEAL website). Business associations, such as CCCMC (outbound mining guidance), use ISEAL's work as a reference in the development of their own standards (CCCMC website). In 2016, the ISEAL Alliance introduced a Chinese version of its code at a forum conducted with Chinese accreditation organizations CCAI and CNCA.

The second approach is illustrated by the evolution of China forestry standards and certification in a process that might be called a nesting of global nonstate actor standards into the Chinese community.

The Forest Stewardship Council (FSC), founded in 1993, is a Bonn (Germany)-based international multistakeholder membership NGO (FSC website). It develops standards and provides for voluntary forest management and chain-of-custody certification. FSC issued its first FSC chain-of-custody certificate in China in 1999. In 2007, WWF set up an FSC China working group with representatives from government, universities, enterprises, and NGOs (WWF website: FSC).

However, the China Forestry Certification Council, a China business association, also has developed standards. The Council, established in 2010, is a member of the Program for Endorsement of Forest Certification Schemes (PEFC), another global forestry nonstate regulator. In 2014, the Council's China Forest Certification Scheme (CFCS) was endorsed by PEFC and embraced by the China's government (PEFC 2014, 2015).

Stream Three: Enterprises take the lead in developing and implementing the rules. This can involve enterprises acting individually in the form of what is often called corporate social responsibility (CSR), trade associations or alliances to promote good practices, or enterprises (including banks and investment consortia) joining forces in support of codes of conduct (e.g. the Equator Principles). Often, the focus is on initiatives aimed at greening supply chains or enhancing their sustainability.

Stream Three initiatives include both those with global connections and those domestically rooted. In addition, as noted above, Chinese business associations play a prominent role in developing social responsibility, including environmental guidelines and standards, doing so in consultation or cooperation with global nonstate networks.

In 2005, the China National Textile and Apparel Council (CNTAC) established an office of social responsibility. CNTAC launched a voluntary "Corporate Sustainability Compact 9000 for Textile and Apparel

Industry" (CSC9000T). CNTAC terms the standard "China's first real attempt to address the concerns of western-based retailers and brands about how their products were being produced" and the beginning of China's efforts to be a "seller" of sustainability standards to the globe (Fudan workshop PPT).

Other associations which have produced social responsibility documents include the China International Contractors Association (CHINCA) (CHINCA website) and the China Chamber of Commerce of Metals, Minerals and Chemicals Importers and Exporters (CCCMC) (CCCMC website).

China business associations now are developing standards for China enterprises going abroad. For example, the CCCMC, in cooperation with the OECD, developed the "Chinese due diligence guidelines for responsible mineral supply in China" (CCCMC website). The aim, according to the "Project brief," is (OECD website):

> for Chinese mining companies undertaking outbound mining investment, cooperation and trade to strictly 'observe the UN Guiding Principles on Business and Human Rights during the entire life-cycle of the mining project' and to strengthen 'responsibility throughout the extractive industries value chain.'

The *CCCMC Chinese Guidelines for Social Responsibility in Outbound Mining Investment (2014)* (CCCMC website) operationalize the call to "conduct risk-based supply chain due diligence in order to prevent engagement with materials that may have funded or fuelled conflict." The guidelines address human rights-related risks (Type 1) and also "risks associated with serious misconduct in environmental, social and ethical issues" (Type 2). Companies are to develop a code of conduct to meet the guidelines' requirements, and there are provisions for third-party audit and certification.

In addition, following Amnesty International's reporting on exploitation of child labor in cobalt mining in the Congo, CCCMC has been credited with playing a lead role in developing "The Responsible Cobalt Initiative," working with OECD and joined by Apple, Samsung, and other multinational electronics suppliers. Cobalt plays an important role in lithium batteries (Whoriskey and Frankel 2016).

Foreign corporate rooted initiatives also are active in China. Walmart played a key role in creating The Sustainability Consortium (TSC) to "green the supply chain." TSC members and partners include manufacturers, retailers, suppliers, service providers, NGOs, civil society organizations, governmental agencies and academics. TSC "convenes these diverse stakeholders to work collaboratively to build science-based decision tools that address sustainability issues that are materially important throughout a product supply chain and lifecycle." In entering China, TSC's U.S. university partners entered into an arrangement with Nanjing University's School of the Environment as a counterpart.

Finally, CSR is emerging, with environment as a prominent focus of specific initiatives. A 2013 Swedish government survey of CSR in China found that environment (at 47%) ranked second only to economic issues (62%) among CSR priorities (CSR-Asia, website). For example, Sinopec, the largest of China's state-owned oil companies, reports on its website (Sinopec website):

> In 2011, we signed responsibility documents of major pollutants discharge reduction target with the government and delegated the target among our subsidiaries. We also developed the 12th Five-Year Plan for Sinopec environmental protection work, which deploys key environment-related tasks in the future.

Mechanisms of Influence or Levers Nonstate Actors Use to Induce Enterprises to Contribute to Environmental Protection and Sustainability

Nonstate actors lack the authority and power states can use to compel change in conduct. To make a difference in addressing environmental problems, they must put pressure on enterprises' core needs; government licenses or permissions; the availability of resources including human and financial capital and natural resources, or markets including the behavior of consumers.

The global repertoire available to nonstate actors to affect these pressure points includes several core mechanisms of influence used in China.

First, in China today, the "shadow of government" occupies a prime role. This is so by virtue of the core role of government both in affecting enterprise CSR conduct and in relation to NGOs and associations, actors whose continued licenses to operate are critically dependent on government.

For example, a presentation by *Syntao*, a China CSR advisory firm whose clients include Chinese and foreign multinationals such as Volkswagen, Coca Cola, Sony, and Lenovo, advises that in China today (Fudan workshop PPT):

- Government has dominant influence;
- Less influence from Investors;
- Less influence from Consumers;
- NGOs are limited in capacity and resources;
- Lack of Trust.

Similarly, the 2013 report by CSR/Asia and the Swedish Embassy summarizes: "A majority of the respondents view the government as the key driver of CSR development in the country" (CSR-Asia website):

> As exemplified by IPE's indices, combining the shadow of government and carrots constitutes a favored approach. The PITI index focuses on encouraging local/local government competition to be transparent. In what might be called "borrowing" government authority, IPE collects and reposts information published by government, with consequent effects on enterprises within localities. IPE's work includes inquiries to enterprises to see if unlawful conduct has been addressed.

As commentators have long noted, modern PRC government is not a monolith (Lieberthal and Lampton 1992; Saiach 2000). Thus, there are opportunities for those pushing change to invoke government attention, but no certainty that attention will be gained, or the intended goal fulfilled.

Second, crisis is a driver in generating and focusing attention that gets government and thereby enterprises to act.

China leader's attention to environment has been continually stimulated by public crises. These include the 2005 Songhua River water spill, which was followed by the elevation of the environmental agency to ministry status, and the 2008 baby milk powder poisoning, which led to a food safety law. Rising public open information-based awareness of PM 2.5 air pollution and the early 2013 Beijing "airpocalypse" were followed by the September 2013 air pollution "action plan" (Young et al. 2015).

Crises similarly provide opportunities for nonstate initiatives. As noted above, a report on child labor in cobalt mining by Amnesty International led to the China mining association's (CCCMC) entrance into the issue. IPE's success in focusing on supply chains has followed IPE's work in bringing to global attention the poisoning of 137 Chinese workers in Apple's supply chain. As a nonstate actor participant at the Fudan workshop put it, Greenpeace investigations play an essential role in providing an opening for organizations that do not play such an adversarial role (Fudan workshop discussion).

Third, there is provision of resources essential to enterprise activity customers, including investors, markets, and insurers that may affect a corporation's reputation.

China's government has experimented with the use of resource incentives for green performance. These include "green credit" (favorable bank treatment for environmental performance) and green government procurement (Qiao and Wang 2011). Currently, "green finance" is a government focus (UNEP/PSI website) and there is study of "sustainable insurance" (Bacani 2015).

In China, as elsewhere, multinational supply chain leverage is a core strategy. From the vantage point of the first stream, IPE's evolving real estate project illustrates an effort to affect a domestic supply chain, seeking to tie China's domestic construction work to compliance with environmental law. IPE engages major China realtors who commit to buy from suppliers who comply with environmental law, as shown by government data. From the vantage of multinational purchasers (the third stream), The Sustainability Consortium's "toolkits" provide key performance indicators (KPIs) for best practices, products, and services. In principle, suppliers and multinational purchasers will use these toolkits in selecting suppliers (The Sustainability Consortium website).

Fourth, there is the provision of expertise that companies value.

As noted above, global second stream nonstate actors, such as the ISEAL Alliance and its members, provide expert services to China business associations, government agencies, and NGO counterparts engaged in developing standards. In addition, they provide expertise directly to enterprises.

For example, the Better Cotton Initiative (BCI) works with China's farmers in "learning groups" that, in turn, are organized into "producing units." The 2014 "Harvest Report" indicates that in China BCI's Implementing Partners worked with 7,028 farmers organized into 10 Producer Units and 9 large farms; 6,458 farmers in China earned a Better Cotton license (Better Cotton Initiative 2014).

The Alliance for Water Stewardship (AWS) similarly views itself as a purveyor of expertise, helping identify available technologies and practices. AWS works across industries (e.g., fashion and micro-electronics) providing understanding of the role of water in the supply chain and supply chain "risks and resilience." AWS's working hypothesis is that adoption of a stewardship approach in managing water by a critical mass of leaders will produce a "tipping point" for an industry sector (AWS Fudan workshop presentation).

Fifth, there is the push and pull of going abroad.

In addition to their roles in multinational supply chains, Chinese enterprises, including major state-owned enterprises, have growing investments and interests in doing business abroad. Thus, as in the CCCMC mining standards and forestry association cooperation with FSC and PEFC, there is interest in being perceived as good global citizens. This role is likely to become more important as China's leaders pursue the Belt and Road Initiative vision (Economist Intelligence Unit 2017).

Sixth, there is what might be called cross-border cultural arbitrage.

Cultural differences can provide both constraints and synergies for nonstate actor efforts to work across borders.

For example, AWS has a substantial presence in Australia, where there is green awareness surrounding fashion brands and consumers. AWS's Australian Fashion Brand Initiative encompasses supply chain links from cotton production to processing to retail sales. The Australian focus provides leverage in China, with China portions of fashion supply chains. In China, AWS works with WWF in cooperation with H&M (AFWS Fudan workshop presentation). (Fudan workshop presentation and discussion.)

Evaluating Performance: The Results of Nonstate Actor Efforts

How effective are the efforts of nonstate actors to protect the environment and promote sustainability in China and what are the determinants of success or failure in this realm? To ask about effectiveness is to launch an effort to answer two types of questions: (i) do the activities of specific nonstate actors succeed in solving the problems or achieving the goals that motivate their efforts? and (ii) do the activities of nonstate actors taken together lead to larger systemic changes in environmental governance in China?

Problem-Solving/Goal Fulfillment

Typically, nonstate actors become engaged in environmental issues because they want to make a difference in solving particular problems or fulfilling more or less well-defined goals. For example, they may be motivated by a desire to clean up polluted water bodies, reduce emissions of greenhouse gases, limit the destruction of forests, or improve the sustainability of production processes in specific industries. Assessing effectiveness, in this context, is a matter of evaluating progress toward the fulfillment of these goals and determining the extent to which a given nonstate actor has played a role in these achievements.

In the general literature on the effectiveness of governance systems, it is common to differentiate among outputs, outcomes, and impacts.

Outputs are the first step; they may take the form of establishing standards, making decisions about awarding certificates, or setting up collaborative arrangements. In our context, the issuance of FSC certificates and the creation of the TSC are prominent examples. Outcomes, by contrast, have to do with measureable changes in behavior. Do corporations that hold FSC certificates make significant changes in their behavior or practices as a result of certification? Impacts bring us to the final and most important stage in the causal chain. Are the problems that lead to action on the part of nonstate actors solved? Are there, for example, measurable improvements in water quality in targeted lakes and rivers? Can we demonstrate that the actions of nonstate actors play significant roles in achieving these results?

In the final analysis, we are interested in problem-solving. Outputs and outcomes are important. But they are steps taken in efforts to solve problems, not ends in themselves. Nevertheless, it is typically easier to document outcomes and especially outputs in contrast to impacts. If the FSC issues a certificate, there is no question about the causal connection. Moving to outcomes already presents problems of demonstrating causality.

At this stage, we lack systematic evidence that would allow us to make solid, data-based assessments of the roles that nonstate actors play in the pursuit of environmental protection in China. Clearly, this is a major challenge for research on the roles of nonstate actors going forward. What we can do is to provide some prominent illustrations or examples from each of the streams of activities that will help to concretize the issues at stake and offer some initial insights regarding the determinants of success in this realm.

For the first stream, the work of IPE and environmental NGOs representing the public interest in court under the environmental law as amended in 2014 may illustrate successes and also, no less useful, highlight the distance still to go and steps that may get there. IPE has pioneered in the use of public information to push compliance with open information laws and the rectification of illegal or harmful conduct it reveals. IPE's annual PITI and CITI reports indicate year-on-year improvements or declines in locality and corporate information reporting. For example, IPE reports that the "top seven" cities now score over 70 points on the PITI index, showing that advanced regions have improved their means for information disclosure. One might hypothesize that IPE's work has an effect. At the same time, however, IPE reports that scores in some regions remain low, and there is insufficient social leverage of the IPE reported data (Fudan workshop Presentation).

In 2015, the first year of environmental "public interest" lawsuits under the 2014 amendment, the initiation of about fifty lawsuits may be treated as progress. Not surprisingly, there were varying degrees of success with regard to outcomes. In some cases, court orders for relief were granted. In other cases, relief was denied or courts did not accept the cases. Perhaps of greatest value from the vantage point of systemic long-term success, the first years highlight challenges.

While an estimated 300–700 NGOs might qualify to bring cases (UNDP 2015), the bulk of the cases were brought by a small number of Beijing-based ENGOs: the China Biodiversity Conservation and Green Development Foundation (CBCGDF) (*Lufahui*), Friends of Nature (FON), and the All China Environment Federation. These organizations are critical but, when compared with U.S. and EU counterparts, Chinese NGOs (*min ban fei qi ye* or *she hui tuanti*) have limited in-house legal and technical staff. The pressure on resources was highlighted by court orders that a successful ENGO plaintiff must still bear its own substantial legal fees, and that unsuccessful plaintiffs may be further burdened with court costs (CBCGDF 2017; Zhang and Tang 2017). Moreover, given limited resources together with the complexity and scale of China's environmental problems, there is a premium on the development of an economy of scale strategy. There are hundreds of large cities and thousands of villages. What assurance is there that limited resources are devoted to highest priorities or that success in one case in one locale will be replicated elsewhere? UNDP has been trying to address the challenges of expertise and scale by working with CBCGDF and other ENGOs to develop a web platform to provide for the accumulation and sharing of case experience regarding the linkage of local citizen complaints and Beijing-based environmental law NGOs, thereby permitting the focusing of scarce resources (UNDP 2016).

In addition, the question of the impacts of court "successes" has yet to be addressed. Prior to the 2014 amendment, there had been successful lawsuits brought by environmental NGOs (Stern 2013; Guttman and Xia 2014). However, there has been limited reporting on whether court victories (whether by order or mediation or settlement) have been put into effect and, if so, what ultimate impact on the environment and lives of citizens they have. Here, too, a UNDP China ENGO project has focused on the potential role of ENGOs in implementing court orders (Id.).

The efforts of the ISEAL Alliance and its members may be used to illustrate stream two, and do so in two respects. First, ISEAL is, by attestation of China trade associations as well as ISEAL Alliance members, playing a role as capacity builder and point of reference in the development of Chinese native standards. Success may be measured by the numbers of Chinese organizations that draw on its work, and the extent to which this reliance is embodied in their further standards and implementation. Second, ISEAL Alliance's effect may be measured by the extent to which ISEAL members adopt and effectuate the Alliance's principles in their China activities. Here, the story appears to be a work in progress.

Many nonstate actors discussed in this chapter are ISEAL Alliance members and publish periodic evaluations in light of Alliance standards. For example, the 2015 Sustainable Agricultural Network (SAN)/Rainforest Alliance Impacts Report: "Evaluating the Effects of the SAN/Rainforest Alliance Certification System on Farms, People, and the Environment" describes the monitoring and evaluation system and provides analyses that focus by crop (bananas, coffee, tea, cocoa) and issue (livelihoods, water, conservation, biodiversity, smart agriculture, and climate change).

However, while reports provide data and success stories for work done outside of China, Chinese analyses are generally limited. For example, the 2015 SAN/Rainforest Alliance Impacts Report provides qualitative analysis and stories of successes regarding both crops and issues. The report shows that RA work is focused on tea, notes that tea originated in China, and that China, Kenya, and India are prime producers today. But the report notes Kenya and India provide over 60% of Rainforest Alliance certified tea production, with China included among 3.5% in the "other country" category. The report reviews evaluations of Kenya and India and "spotlights" a Kenya farmer. But there is yet to be similar in-depth reporting on China.

ISEAL Alliance members do publicly report on a variety of China's output measures. These include "softer" measures, such as Memoranda of understanding, and numbers of members, workshops, conferences, trainings, or other awareness-raising activities. For example, the Marine Stewardship Council reports, in cooperation with the China Chain Store and Franchise Association, WWF, and UNEP, on a "sustainable seafood week" (Marine Stewardship Council website).

The measures also include indicators regarding application of their standards through certification. In 2016, for example, FSC reports 4,000 chain-of-custody certificates in China, mostly for export and hectares of forest under certification (FSC website). Similarly, in 2016, the Rainforest Alliance reports on China activities included (Rainforest Alliance 2015, Fudan workshop PPT): (i) 36 farms and forest management units under sustainable management, (ii) 8 international buyers and hundreds of suppliers engaged in responsible sourcing, and (iii) 15,000 individuals trained in sustainable land management.

In some cases, public reports go further toward discussions of outcomes and impacts. For example, the Better Cotton Initiative (BCI) reports its China activities as a component of annual "harvest reports." BCI reports continued year-on-year improvements. The 2014 data include numbers of farmers, areas under cultivation, and yields. The reports compare BCI farmers with "comparison" farmers on indicators including yield, synthetic and organic fertilizer use, pesticide and water use as well as profits (Better Cotton Alliance 2014). The 2014 Harvest Report documents that in all cases the BCI licensees did better than the comparison group.

Next, there are the corporate-driven activities of stream three. Here, China offers three models: activities driven by multinationals such as The Sustainability Consortium (TSC), activities driven by Chinese business associations, and CSR.

The TSC "2016 Impact Report" provides aggregate data (and "scores") by product group and lifecycle stage, but it does not contain country-specific data (The Sustainability Consortium website).

Chinese companies increasingly provide CSR reports with "environment" or "sustainability" components. In relation to CSR, The Sino-Sweden China CSR website includes an environmental "best practices" page with summary case studies identifying practices and "achievements" (Sino-Sweden CSR, website).

Systemic Effects

Whatever conclusions we reach about the effectiveness of the efforts of individual nonstate actors in addressing particular environmental issues in China, it is pertinent also to ask whether the efforts of these actors taken together are bringing about significant changes in China's institutionalized environmental governance processes. Many specific initiatives may fall short or fail altogether; many actors may end up with disappointing records regarding efforts to solve major environmental problems. Even so, it is possible that the environmental movement spearheaded by an array of nonstate actors may be playing a role in bringing about systemic changes in the ways environmental problems are handled in China.

It is helpful to introduce a distinction between two types of changes in thinking about these systemic effects: (i) changes in governmental structures and processes relevant to addressing environmental issues in China and (ii) changes in the scope of opportunities for nonstate actors to play effective roles in environmental governance going forward.

As in the case of problem-solving, we are not currently in a position to produce data-based conclusions about these matters. That will require a sustained effort to collect and evaluate data over a period of time. But we can offer some initial observations about the systemic effects of the efforts of nonstate actors in both domains.

Structural changes to watch include: (1) the developing role of China trade associations in setting standards for both domestic and "going abroad" China enterprises, and the related development of CSR, (2) the role of IPE in developing "NGO driven" information-based regulation and environmental NGOs in bringing "public interest" court cases, and

(3) the iterative effect, if any, between China "outbound" and domestic standard setting and implementation.

In comparison with western spawned nonstate arrangements, structural developments in China feature core roles for public service units (*shiye danwei*) and business associations with close government connections, and, to date, more limited development of independent NGOs, whether measured in number or capacity of NGOs like IPE or the growth of NGOs making use of environmental law.

In the context of the scale and complexity of China's environmental challenges, the evolving structure has advantages and disadvantages. Given the small size of the central government's civil service environmental workforce (500 or fewer civil servants in the central Ministry of Ecology and Environment), China's business associations may provide potential nationwide reach into the conduct of Chinese enterprises, along with access to data and information about their conduct that critically supplements the influence of government and organizations independent of enterprises. In addition, the role of trade associations in developing "voluntary" standards may provide a reasonable testing ground for "policy learning" in the development of China's standards.

On the other hand, public service units may generate routinized products and business associations may provide lowest common denominator standards and oversight, so that the role NGOs (independent of business associations and enterprises) play elsewhere in providing stimulus and oversight may be limited in China.

Interaction among elements of the evolving institutional structure may be key. China's government is mandating CSR as well as supporting the activities of trade associations. How will the CSR efforts of individual enterprises, presumably motivated at least in part by a desire to portray themselves as leaders and distinguish themselves from competitors, relate to associational standard setting? Will CSR leaders provide stimulus for innovation initiating "races to the top" that NGOs lead elsewhere? As a corollary, in China's remarkable development, local governments have played key roles in stimulating "races to the top" for GDP growth. In relation to measures like the performance indicators being developed by IPE and potential litigation, will local governments provide leadership in promoting green races to the top?

Finally, there is the past and present role of "foreign NGOs" in the current structure. Foreign NGOs have played multiple roles: (1) providing expertise and capacity building and financial support for Chinese counterparts in standard development and implementation, (2) providing cross-border leverage, as in the case of AWS's capacity to bring Australian fashion interests to bear on China fashion supply chain, and (3) providing investigative reports and campaigns, as with Greenpeace and Amnesty International, that provide space for foreign and domestic actors. As China develops its own expertise and institutions, will the need for foreign expertise decline? Will the implementation of the 2016 Foreign NGO law produce major changes?

Revisiting the Western Literature: Lessons from the Chinese Experience

What are the implications of developments in China for our general understanding regarding the roles of nonstate actors in addressing environmental issues?

We reemphasize at the outset that the study of Chinese governance processes indicates that the concept of "nonstate actor" (or terms like "NGO" or "private authority") is not robust enough to capture the reality of China's emerging efforts to deal with environmental issues. This challenge of vocabulary points to a matter of substance, not simply form. Government (again including the Party) plays a different role in China today than it does in locales, such as the U.S., EU, and Commonwealth countries, where the notion of "nonstate actor" has evolved in the environmental context.

While Chinese and western actors use similar tools (e.g., information disclosure, certification, and accreditation), these tools operate differently in the different systems. In China, as we have discussed, the role of "the shadow of government" as a mechanism of influence appears to be far more consequential than, for example, the role of markets, financial institutions, or consumers.

For scholars and practitioners alike, there is value to a vocabulary that is sufficient to encompass western "nonstate actors" and Chinese counterpart systems. Further analysis should help researchers to clarify the meaning(s) and value of the "nonstate actor" concept—not only in China but also in the West—and practitioners to understand how cross-border arrangements need to be adjusted for application to "local" conditions.

With an appropriate vocabulary in hand, there will then be opportunities for comparisons that may lead to new scholarly insights and practical learning. Most basically, how effective will the governance processes developing in China with their emphasis on actors like government-related public service units and business associations be compared with western models emphasizing the role of more autonomous nonstate actors? Next, and no less basically, how will Chinese processes play out in the context of China's dual role as one of the world's largest economies and as a developing economy? This perspective suggests a number of substantive points of comparison.

First, China and many Chinese enterprises have interests in striving to meet top global standards and in cooperating with global nonstate actor environmental standard setters toward this end. Thus, Chinese business association standard developers work with and reference the work of the ISEAL Alliance and its members. As exemplified by the CCCMC work on minerals, China's mineral trade association is working with the OECD and global suppliers on global responsibility guidelines, and also addressing child labor in cobalt production. The guidelines for Chinese enterprises going abroad work off global standards. The evolution and implementation of these guidelines in light of China leadership's current Belt and Road Initiative vision should be a particular point of interest and focus.

As a corollary, there is the twenty-first-century question of China as a leading global consumer of multinational products as well as a world leading supply chain source. For example, China is not a significant palm oil producer but, with India, it is the world's largest importer (IISD 2014). In May 2013, the Roundtable for Sustainable Palm Oil (RSPO) signed an MOU with the China Chamber of Commerce for Import and Export of Foodstuffs, Native Produce and Animal By-Products (CFNA). At the 2016 third China palm oil summit, RSPO announced over 50 Chinese members. These include Sinograin, a large state-owned grain company, and the China National Cereals Oils and Foodstuffs Corporation (COFCO), China's largest palm oil importer, with two palm facilities certified (Kingdom of Netherlands website). As China's leaders seek to draw on increased citizen purchasing power to transform the economy, it is a common complaint that citizens prefer to spend money abroad or on foreign products. How will Chinese enterprises on the consuming end of global supply chains drive global standards? Will evolving Chinese practices provide powerful pushes or drags on the evolving global system of environmental protection?

Second, and closely related to the previous point, is the relationship between environmental and other standards in China's governance processes. Global nonstate actors may integrate environmental standards with labor, community, or human rights standards. For example, the SAN/Rainforest Alliance Report focuses on worker conditions as an issue, using India as a case. China is not discussed. The difficulties of nonstate actor efforts to engage in labor and environmental standard setting in China are often reported. At the same time, Chinese government and *shiye danwei* CSR guidelines follow "western" guidelines in the inclusion of factors beyond the environment. In the work of the China Chamber of Commerce of Metals Minerals and Chemicals Importers and Exporters (CCCMC), for example, labor and human rights are an express component of "going abroad" standard development. Attention to the balance between the language of China's nonstate actor standard setting and the on the ground actuality will be a subject calling for continued attention.

Third, a core variable in nonstate actor efforts is the understanding of the "corporate suite black box" or how to develop "corporate suite" awareness that will provide receptivity to, and incentives for, cooperation and embracing of standards. Here, again, the Chinese system provides a basis for comparison. Elsewhere, inter-industry competition itself (including competition for consumers or financial sources) might be key. In China, the "shadow of government" remains the key. Thus, the 2013 CSR-Asia/Swedish Embassy China CSR review (op. cit. at 24) found a "common opinion" that:

> while there is a solid framework of environmental legislation in place, it is poorly enforced. [Survey results] highlight insufficient monitoring of regulatory compliance as the key obstacle preventing CSR implementation in China ...According to some respondents, even among those companies that do comply with regulations, environmental responsibility is largely perceived as a passive obligation to meet the requirements of regulators or clients.

There are multiple stages of corporate engagement. Initially, corporate focus is on compliance with laws, with nonstate actor leverage focused on law compliance. At a second stage, the focus is on potential "win/win" efficiencies to be gained by using new technologies or methods in ongoing operations. At a third stage, the focus includes looking at the supply chain as a whole with potential focus on rethinking products or services. How will China CSR evolve, and how rapidly?

Fourth, there are questions regarding the continued role of "foreign NGOs" in environmental governance in China. The implementation of the 2016 "Foreign NGO law" will put the role of foreign NGOs to the test. China's remarkable development has placed high and practical value on working with global nonstate actor environmental regulators. But questions abound about the new law's implementation. These include the readiness of approved government entities to take on sponsorship of a foreign NGO, particularly given the limited numbers and work commitments of China's civil servants. In the absence of high-level support, the question arises of whether the risks of responsibility outweigh the benefits. They also include, from the "foreign NGO" perspective, the reporting requirements to the Public Security Ministry and the risks to local (as well as foreign) staff and organizations from non-adherence to rules. Finally, they include the capacity of foreign NGOs, such as Greenpeace, that may be said to play a keystone species role in the nonstate actor landscape through research and reports that help catalyze opportunities for others.

A further question is what alternatives to registration under the law may be available, with what effective levels of risk for the foreign organization and tolerance by China's government. These include, for example, drawing on the role of many foreign nonstate actors as advisors to enterprises or seeking status as a business or business consultancy rather than an "NGO."

China's many environmental challenges are well known. Now, as China assumes a leading role in global environmental affairs, it is essential for scholars and practitioners to improve understanding of how the roles pioneered by western "nonstate actors" are being translated into a China context in which China counterpart institutions are not only important domestically but also may assume global leadership roles.

References

Auld, Graeme. 2014. *Constructing Private Governance: The Rise and Evolution of Forest Coffee and Fisheries Certification.* New Haven: Yale University Press.

Bacani, Butch. 2015. A Systematic View of the Insurance Industry and Sustainable Development: International Developments and Implications for China. https://unepinquiry.org/wp-content/uploads/2015/10/greening-chinas-financial-system-chapter-9.pdf.

Barboza, David. 2011. Workers Sickened at Apple Supplier in China. *New York Times*, February 22.

Baron, David P. 2009. *Business and Its Environment*, 6th ed. Prentice Hall.

Bartley, Tim, Sebastian Koos, Hiram Samel, Gustavo Setrini, and Nik Summers. 2015. *Looking Beyond the Label: Global Industries and the Conscientious Consumer*. Indiana University Press.

Better Cotton Initiative. 2014. Better Cotton Initiative 2014 Harvest Report. https://bettercotton.org/wp-content/uploads/2013/12/FINAL-HARVEST-REPORT-2014-updated-2pg1.pdf.

BSD Consulting. 2016. Sustainability Reporting Standards in China. https://www.bsdconsulting.com/insights/article/sustainability-reporting-standards-in-china.

Buthe, Tim, and Walter Mattli. 2011. *The New Global Rulers: The Privatization of Regulation in the World Economy*. Princeton University Press.

Cashore, Benjamin, Graeme Auld, and Deanna Newsom. 2004. *Governing Through Markets Forest Certification and the Emergence of Non-state Authority*. New Haven: Yale University Press.

CBCGDF (China Biodiversity Conservation and Green Development Foundation). 2017. Look Back on the Changzhou Environment Public Interest Case. February 7. https://cbcgdf.wordpress.com/2017/02/07/look-back-on-the-changzhou-environmental-pubic-interest-litigation-case/.

China Daily. 2012. Number of NGOs in China Grows to Nearly 500,000. March 20. https://www.chinadaily.com.cn/china/2012-03/20/content_14875389.htm.

China Chamber of Commerce of Metals, Minerals, and Chemicals Importers and Exporters (CCCMC) website. https://en.cccmc.org.cn/.

China Chamber of Commerce of Metals, Minerals, and Chemicals Importers and Exporters (CCCMC). 2015a. Chinese Due Diligence Guidelines for Responsible Mineral Supply Chains Project Brief. May. https://www.oecd.org/daf/inv/mne/CCCMC-Guidelines-Project%20Brief%20-%20EN.pdf.

China Chamber of Commerce of Metals, Minerals, and Chemicals Importers and Exporters (CCCMC). 2015b. "Chinese Due Diligence Guidelines for Responsible Mineral Supply Chains." https://www.cccmc.org.cn/docs/2016-05/20160503161408153738.pdf.

China International Contractors Association (CHINCA) website. https://www.chinca.org/.

China Ministry of Civil Affairs (MCA) website. https://www.mca.gov.cn/article/sj/tjgb/201607/20160700001136.shtml.

China National Accreditation Service for Conformity Assessment (CANS). https://www.cnas.org.cn/english/introduction/12/718683.shtml.

China National Institute of Standardization (CNIS) website. https://en.cnis.gov.cn/bzygk/kyly/.

China National Offshore Oil Corporation (CNOOC) website. https://www.cnoocltd.com/col/col7601/index.html.

China National Textile and Apparel Council (CNTAC) website. https://english.ctei.cn/.

China Petroleum and Chemical Corporation (Sinopec). Social Responsibility webpage. https://www.sinopecgroup.com/group/en/socialresponsibility/.

Chinese Standards Shop. *Guidance on Social Responsibility Reporting*. GB/T 36001-2015. https://www.chinesestandard.net/default.aspx?PDF-English-ID=GB/T%2036001-2015.

Communist Party of China (CPC). 2016. *The 13th Five-Year Plan for Economic and Social Development of the People's Republic of China (2016–2020)*.

Compilation and Translation Bureau, Central Committee of the Communist Party of China. Beijing, China: Central Compilation and Translation Press. https://en.ndrc.gov.cn/newsrelease/201612/P020161207645765233498.pdf.

CSR-Asia, Embassy of Sweden Beijing. "A Study on Corporate Social Responsibility Trends in China." https://www.csr-asia.com/report/CSR-development-and-trends-in-China-FINAL-hires.pdf.

Draft Amendment to the China National Standardization Law. https://hope.haier.com/nasi/f/standardNews/detail/105.

Economist Intelligence Unit. 2017. One Belt one Road: An Economic Road Map. *The Economist*. https://www.eiu.com/public/topical_report.aspx?campaignid=OBORSept2016.

Environmental Protection Law of the People's Republic of China. https://www.chinadialogue.net/Environmental-Protection-Law-2014-eversion.pdf.

FSC (Forestry Stewardship Council) websites (Chinese and English). https://cn.fsc.org/cn-cn ; https://ic.fsc.org/en.

FSC website. https://ic.fsc.org/en/news-updates/id/1548.

Fürst, Kathinka. 2016. *Regulating through Leverage: Civil Regulation in China*. PhD Thesis, Amsterdam University Press.

Global Reporting Initiative: Linking Cass CSR-3 and GRI's G4 Sustainability Reporting Guidelines: 2014. https://www.globalreporting.org/resourcelibrary/Linking-CASS-CSR-3.0-and-GRI's-G4-Sustainability%20Reporting-Guidelines.pdf.

Green, Jessica F. 2014. *Rethinking Private Authority: Agents and Entrepreneurs in Global Environmental Governance*. Princeton University Press.

Guttman, Dan. 2000. Public Purpose and Private Service: The Twentieth Century Culture of Contracting Out and the Evolving Law of Diffused Sovereignty. *Administrative Law Review* 52 (3): 859–926.

Guttman, Dan. 2015. The Global Vernacular of Governance and Instruments: Translating Between the Operating Systems of China and the United States. *Asia Pacific Journal of Public Administration* 37 (4): 253–263.

Guttman, Dan, Yaqin Song, and Haiming Li. 2013. U.S. Government Contracting and China Shi Ye Dan Wei-Two Shadow Governments: Path Dependency from Opposite Directions, or Mutual Learning. *Asia Pacific Journal of Public Management*. 35 (1): 1–51.

Guttman, Dan, and Kunbao Xia. 2014. UNDEF 4th Round Project: "Protect the Environmental Rights and Justice of the Public". https://www.un.org/democracyfund/sites/www.un.org.democracyfund/files/UDF-CPR-09-320_ER.pdf.

Hong Kong Stock Exchange Environmental. n.d. *Social and Governance ("ESG") Guidelines*. https://www.hkex.com.hk/eng/rulesreg/listrules/listsptop/esg/.

Hsueh, Roselyn. 2011. *China's Regulatory State a New Strategy for Globalization*. Cornell University Press.

International Institute for Sustainable Development (IISD). 2014. Meeting China's Global Resource Needs. Managing Sustainability Impacts to Ensure Security of Supply. Palm Oil Pilot Study. February. https://www.iisd.org/sites/default/files/publications/china_global_resource_palm_oil.pdf.

International Trade Administration (ITA) website. https://www.trade.gov/td/standards/markets/East%20Asia%20Pacific/China/China.pdf.

Institute of Public and Environmental Affairs (IPE) website https://wwwen.ipe.org.cn/.

IPE. 2015. Greening the Global Supply Chain: CITI Index. https://www.esquel.com/download/CITI-report-2015-EN.pdf.

ISEAL Alliance website: https://www.isealalliance.org/our-members/full-members.

ISEAL Alliance. 2014. ISEAL Code of Good Practice for Assessing the Impacts of Social and Environmental Standards Systems (6.0 version). December. https://www.isealalliance.org/sites/default/files/ISEAL%20Standard%20Setting%20Code%20v6%20Dec%202014.pdf.

Jing, Yijia. 2015. Between Control and Empowerment: Governmental Strategies of Nonprofit Development in China. *Asian Studies Review* 39: 589–608.

Kingdom of the Netherlands website. The Third China Roundtable on Palm Oil. https://www.netherlandsworldwide.nl/countries/china/news/2016/08/04/the-third-china-roundtable-on-sustainable-palm-oil.

Koleski, Katherine. 2017. The 13th Five Year Plan. *U.S./China Economic and Security Review Commission*, February 14. https://www.uscc.gov/sites/default/files/Research/The%2013th%20Five-Year%20Plan.pdf.

Lewis, Mark Edward. 2007. *The Early Chinese Empires: Qin and Han*. Harvard University Press.

Lieberthal, Kenneth G., and David M. Lampton, eds. 1992. *Bureaucracy, Politics and Decision Making in Post-Mao China*. Berkeley: University of California Press.

Li, Dun (ed.). 2016. *2015 Review of Public Interest Litigation in Environmental Protection* (环境公益诉讼起诉观察报告2015年卷). China: Law Press.

Marine Stewardship Council website. https://www.msc.org/newsroom/news/chinese-retailers-support-msc-sustainable-seafood-week.

Milder, Jeffrey C., and Deanna Newsom. 2015. SAN/Rainforest Alliance Impacts Report 2015. *Rainforest Alliance/ SAN* (December 17): 56–67. https://www.rainforest-alliance.org/impact-studies/impacts-report-2015.

National Group Standards Information Platform. 2016. *Suggestion on Development of Association Standards* (guanyu peiyu fazhan tuanti biaozhun de zhidao yijian). https://www.ttbz.org.cn/zcwj/zcwj2/201603/20160317/j_201603171400310004894.html.

OECD website. 2014. OECD/CCCMC Memorandum of Understanding. https://www.oecd.org/daf/inv/mne/MoU-CCCMC-OECD.pdf=MOU.

PEFC. 2014. China's National Forest Certification System Achieves PEFC Endorsement. March 3. https://pefc.org/news-a-media/general-sfm-news/1459-china-s-national-forest-certification-system-achieves-pefc-endorsement.

PEFC. 2015. China Adopts New Forest Certification Regulation. July 3. https://pefc.org/news-a-media/general-sfm-news/1889-china-adopts-new-forest-certification-regulation.

Qiao, Yuhua, and Conghu Wang. 2011. Issues and Challenges in Implementing China's Green Public Program. *Journal of Environmental Protection* 8: 1034–1045.

State Owned Assets Supervision and Administration Commission of the State Council of the Peoples Republic of China (SASAC). 2011. Guidelines to State-Owned Enterprises Directly Under the Central Government on Fulfilling Corporate Social Responsibilities. https://en.sasac.gov.cn/n1408035/c1477196/content.html.

Saiach, Tony. 2000. Negotiating the State: The Development of Social Organizations in China. *The China Quarterly* 161: 124–141.

Simon, Karla W. 2013. *Civil Society in China: The Legal Framework from Ancient Times to the New Reform Era*. Oxford University Press.

Sino-Swedish Corporate Social Responsibility website. https://csr2.mofcom.gov.cn/.

Standardization law of the People's Republic of China. 2017, November 4. https://www.npc.gov.cn/englishnpc/Law/2007-12/12/content_1383927.htm.

State Council (SC) Document. *Circular of the State Council on Distributing the Plan for Deepening Standardization Reforms*. No. [2015] 13. https://sesec.eu/app/uploads/2015/06/2015_04_SESECIII_Newsletter_March_2015_AnnexA_StateCouncilDocumentNo2015-13.pdf.

Stern, Rachel E. 2013. *Environmental Litigation in China: A Study in Political Ambivalence*. Cambridge University Press.

Supreme Court of the People's Republic of China website. https://court.gov.cn/fabu-xiangqing-13025.html.

Tang, Shui-Yan., and Carlos Wing-Hung. Lo. 2009. The Political Economy of Service Organization Reform in China: An Institutional Choice Analysis. *Journal of Public Administration Research and Theory* 19 (4): 731–767.

The Sustainability Consortium. 2016. Greening Global Supply Chains: From Blind Spots to Hot Spots to Action. https://www.sustainabilityconsortium.org/wp-content/2016-impact-report/index.php.

UNDP. 2015. Issue Brief: Environmental Public Interest Litigation in China. May 12. https://issuu.com/undp-china/docs/issue_brief_environemental_public_i.

UNDP. 2016. Making Environmental Law Work: UNDP Co-organises Consultation to Test Proposal for a Platform of Common Knowledge. June 22. https://www.cn.undp.org/content/china/en/home/presscenter/articles/2016/06/22/making-environmental-law-work-undp-co-organises-consultation-to-test-proposal-for-a-platform-for-public-participation-in-environmental-rule-of-law.html.

UNEP/PSI website. n.d. Greening China's Financial System. *PSI and UNEP Finance Initiative*. https://www.unepfi.org/psi/greening-chinas-financial-system/.

Vogel, David. 2006. *The Market for Virtue: The Potential and Limits of Corporate Social Responsibility*. Washington, DC: Brookings Institution Press.

Wang, Ping, and Liang Zheng. 2016. Beyond Government Control of China's Standardization System—History, Current Status and Reform Suggestions. August 1. https://papers.ssrn.com/sol3/papers.cfm?abstract_id=2745621.

Whoriskey, Peter, and Todd C. Frankel. 2016. Tech Giants Pledge to Keep the Children Out of Cobalt Mines That Supply Children. *The Washington Post*, December 20. https://www.washingtonpost.com/news/the-switch/wp/2016/12/20/tech-giants-pledge-to-keep-children-out-of-cobalt-mines-that-supply-smartphone-and-electric-car-batteries/?utm_term=.08e142f1855c.

WWF website. https://en.wwfchina.org/en/what_we_do/forest/sustainable_forest_management/.

WWF. 2010. Certification and Roundtables Do They Work? WWF Review of Multistakeholder Sustainability Initiatives. September. https://zembla.vara.nl/pdf/wwf_certification_and_roundtables_briefing.pdf.

Young, Oran R., Dan Guttman, Ye. Qi, Kris Bachus, David Belis, Hongguang Cheng, Alvin Lin, et al. 2015. Institutionalized Governance Processes: Comparing Environmental Problem Solving in China and the United States.

Global Environmental Change 31: 163–173. https://doi.org/10.1016/j.gloenvcha.2015.01.010.

Zhang, Chun, and Damin Tang. 2017. The Changzhou Soil Pollution Case If Far from Over. *China Dialogue*, February 24. https://www.chinadialogue.net/article/show/single/en/9630-The-Changzhou-soil-pollution-case-is-far-from-over.

The Landscape of Non-state Actors and China Environmental Governance: Illustrative Roadmaps to Processes and Institutions

Ruoyu Chen and Xingjian Zhang

Introduction

This chapter provides organizational chart landscapes of the roles of non-state actors in China's environmental governance. The charts in this chapter attempt to place specific non-state actors in the context of one

---

Ruoyu Chen is the content author, and Xingjian Zhang is the graphic designer for the charts. The first three sections of this chapter (general overview, standard-making, supervision and certification) are based on previous work by Carmen Chen, Yangyang Cui and Harold Wang from Fudan University. The chapter has also benefited from comments and suggestions from further book chapter authors and outside reviewers with expertise in china environmental law and governance.

R. Chen (✉)
Stern School of Business, New York University, New York, NY, USA
e-mail: rc3494@nyu.edu

X. Zhang
University of Notre Dame, South Bend, IN, USA

D. Guttman et al. (eds.), *Non-state Actors in China and Global Environmental Governance*, Governing China in the 21st Century,
https://doi.org/10.1007/978-981-33-6594-0_3

another and specific government agencies. The charts provide vantages of the landscape from multiple points of view.

The first three narratives and related charts seek to lay out government and non-state institutions involved in core processes of standard setting, certification, supervision and enforcement.

The following narratives and accompanying charts focus more closely on:

The core environmental challenges of air and water pollution;
Textile and air-conditioning governance, as examples of industry sector governance;
Selected further environmental governance initiatives.

As further explained and illustrated in the succeeding chapters, the landscapes include the development and implementation of rules or standards which are required by government, rules or standards which are voluntary, and related support.

We note that the landscapes were prepared in consultation with chapter authors and further experts on China environmental law and governance. However, the landscapes are complex (e.g., some institutions may be inadvertently overlooked) and changing (e.g., through new laws or policies that may alter institutional relationships, government reorganizations, the creation of new non-state actors and/or the revision of institutional missions). Therefore, these landscapes should be seen as snapshots or starting points for use by scholars and practitioners. We hope and expect that further study and practice will provide for their refinement, and, as appropriate, revision or correction.

The Landscape of China Standard-Related Processes: An Overview of Institutional Actors

Figure 1 on the following page is vertically divided into three parts: standard-making, supervision and certification and evaluation.

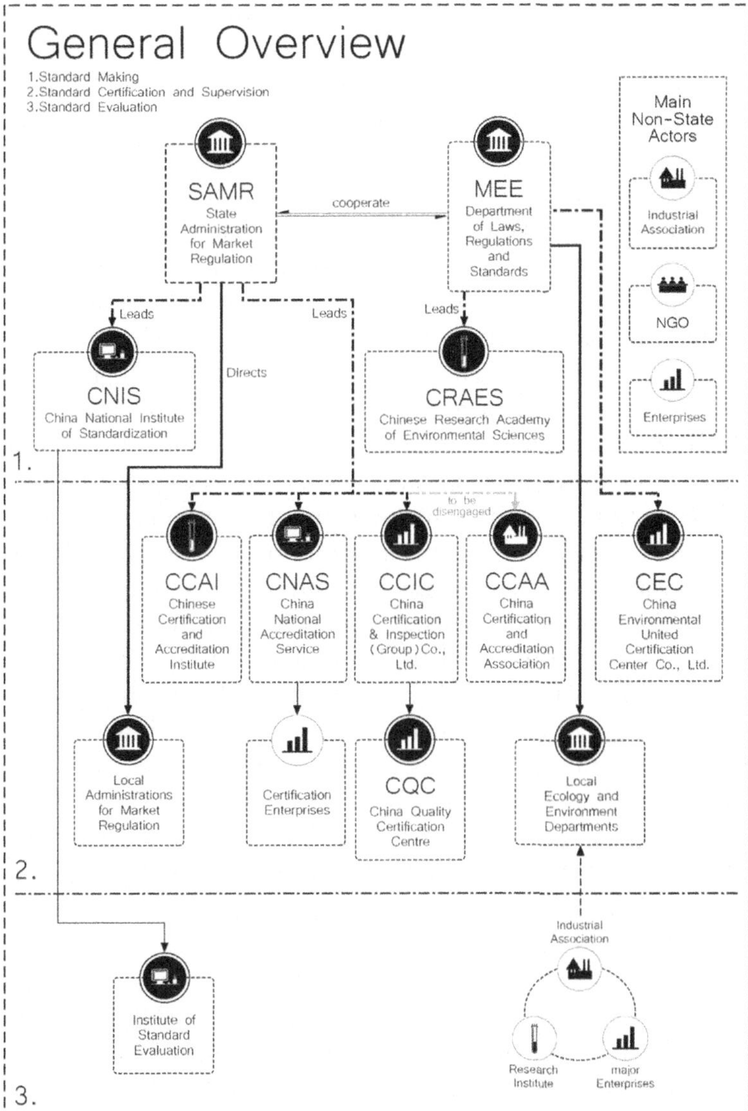

Fig. 1 General overview of institutional actors in China's standard-related processes

The top layer of the chart focuses on institutional actors in government standard setting. (See chapter "Non-industry and Nonstate Actors Contribution in the Standard Drafting Process: Examples from the Development of China Room Air Conditioner Standards" for more on standard-setting law and the standard-setting process.) The Standardization Administration of China (SAC, 国家标准化管理委员会) had central authority. Following 2018 organizational reform, SAC was merged into the new State Administration for Market Regulation (SAMR, 国家市场监管总局). Though the name SAC is still kept after the merger, the SAC essentially became part of the SAMR. Currently, the Standard Technical Management Division (标准技术管理司) and the Standard Innovation Management Division (标准创新管理司) under the SAMR are in charge of standardization issues. The former focuses on standardization strategies, planning, policy and regulation frameworks, as well as national standard-making and law enforcement assistance and the management of technical committees. The latter focuses on industry, local, group, enterprise and international standards. Various other state and non-state actors also participate in standardization. The China National Institute of Standardization (CNIS, 中国标准化研究院), a *shiye danwei* under SAMR, engages in research and policy assistance. In the case of environmental standards, the Ministry of Ecology and Environment (MEE) and its Department of Laws, Regulations and Standards play key roles. MEE is aided by *shiye danwei* under MEE that include the Chinese Research Academy of Environmental Sciences (CRAES, 中国环境科学研究院). The Environmental Standards Institute under CRAES, similar to the CNIS, is involved in the research and actual development of standards.[1]

Depending on the subject matter, other ministries, including the National Development and Reform Commission (NDRC), Ministry of Industry and Information Technology (MIIT), Ministry of Agriculture (MOA), Ministry of Water Resources (MWR), Ministry of Natural Resources (MLR), National Forestry and Grassland Administration (NFGA), may also participate in environmental standard setting.

Depending on subject matter, and as we will see from industry and issue-specific charts, industry associations, enterprises and universities are also involved in standard setting. Technical committees under the

SAMR form another set of standard-setting-relevant institutions. These committees are often housed within industry associations and research institutes.

The middle layer of the chart focuses on institutional actors in supervision and certification. Supervision and law enforcement largely fall under the authority of MEE and local ecology and environment departments and bureaus, while ministries and local departments and bureaus of water resources, natural resources, housing and urban–rural development, agriculture and rural affairs, transport and forestry are also involved for environmental regulation within their jurisdiction.

China's civil society organizations, as discussed in other chapters, play roles in exerting supervisory pressure on enterprises that violate environmental regulation. These include organizations qualified under 2014 amendments to the Environmental Protection Law to go to court in the name of the public interest, and further organizations, such as the Institute of Public and Environmental Affairs (IPE) (see chapter "Extending Enforcement: How the Institute of Public and Environmental Affairs Leverages Public Information to Strengthen Environmental Governance" on IPE).

Certification involves a wider range of actors. The Certification Regulation and Management Division (认证监督管理司) and the Accreditation and Inspection & Testing Regulation and Management Division (认可与检验检测监督管理司) under the SAMR undertake policy formation, fostering and law enforcement assistance related to certification and accreditation. In addition, various *shiye danwei*, state-owned enterprises (*guo you qi ye*) and industry associations (*she hui tuanti)* under the SAMR also play important roles, including the China Certification & Accreditation Association (CCAA, 中国认证认可协会), China National Accreditation Service for Conformity Assessment (CNAS, 中国合格评定国家认可委员会), China Certification and Accreditation Institute (CCAI, 国家市场监督管理总局认证认可技术研究中心) and China Certification & Inspection Group Co., Ltd. (CCIC, 中国检验认证集团), which controls the China Quality Certification Centre (CQC, 国家质量认证中心). These organizations have their own responsibilities, which will be later introduced in the chart on supervision and certification. MEE is also present in the certification field.

The China Environmental United Certification Center Co., Ltd. (CEC, 中环联合认证中心), which is a state-owned company under MEE, is dedicated to providing third-party certification involving environmental protection, energy-saving and climate change.[2]

The bottom layer illustrates actors in standard evaluation. A department under CNIS, the Department of Standard Evaluation/Institute of Standard Evaluation, takes charge of this function. Its range of operations includes evaluation for the development of voluntary national standards, evaluation of industry and regional standards, statistical analysis of standard implementation and research on standard implementation and on the efficiency of technical committees under SAC. Local ecology and environment departments, together with industry associations, research institutes and major enterprises also participate in the evaluating process.

In addition to these specific governance actors, environmental governance also involves national- and local-level planning and policymaking. For example, in relation to the Ministry of Ecology and Environment (MEE), under the CRAES, there are the Environmental Policy and Strategic Environmental Impact Assessment Center and Environmental Information Center. In addition to CRAES, the China Academy of Environmental Planning (CAEP, 环境规划研究院) and the Policy Research Center for Environment and Economy (PRCEE, 环境与经济政策研究中心) are also shi*ye danwei* under the MEE. CAEP also conducts research in more technical issues such as methods for nonpoint source water pollution control, coal-burning pollution control, instruction for medical waste disposal, among many others.[3] In addition, the Institute for Urban and Environmental Studies (IUES, 城市发展与环境研究所) under the China Academy of Social Sciences (CASS, 中国社会科学院) also may engage in research, planning and environmental risk and damage evaluation.

China Standard-Making: An Overview of Institutional Actors and Processes

Figure 2, on the preceding page, illustrates selected elements in the environmental standard-making process in China. Environmental standards in China can be divided into national standards, local standards, industry standards, group standards and enterprise standards, according to China's Standardization Law. (For more detail on the *shehui tuanti* voluntary standard-making process, see chapter "Will China Industrial Organizations Succeed in Addressing the "Trust Deficit" Arising from Global Supply Chain Governance?"; on the government standard-making process, see chapter "Non-industry and Nonstate Actors Contribution in the Standard Drafting Process: Examples from the Development of China Room Air Conditioner Standards".) According to this law, national standards are prioritized; in industries where no national standard yet exists but there is a need for standardization, relevant government agencies under the State Council (including the MEE) can propose to develop industry standards. To satisfy special technical requirements including local natural conditions and social customs and norms, provincial or municipal level government bodies in charge of standardization can propose local standards. National standards can be voluntary or mandatory, whereas industry and local standards are voluntary. Group standards are made by "social groups" (*shehui tuanti*) including business associations to "satisfy the needs of market and innovation" and are voluntary. Enterprise standards are made by one enterprise or a group of enterprises according to their needs. Both group and enterprise standards are made in an autonomous fashion relatively independent from government intervention.

Different types of standards within the environmental governance field may focus on different issues, for example, overall environmental quality standards, enterprise emission standards or sample standards. In the chart, standards are categorized in terms of applicability, and whether they are voluntary or mandatory.[4]

The SAMR and the MEE, as national ministry-level authorities, coordinate, guide and supervise the development of mandatory and voluntary local standards. They work, in turn, with corresponding local agencies.

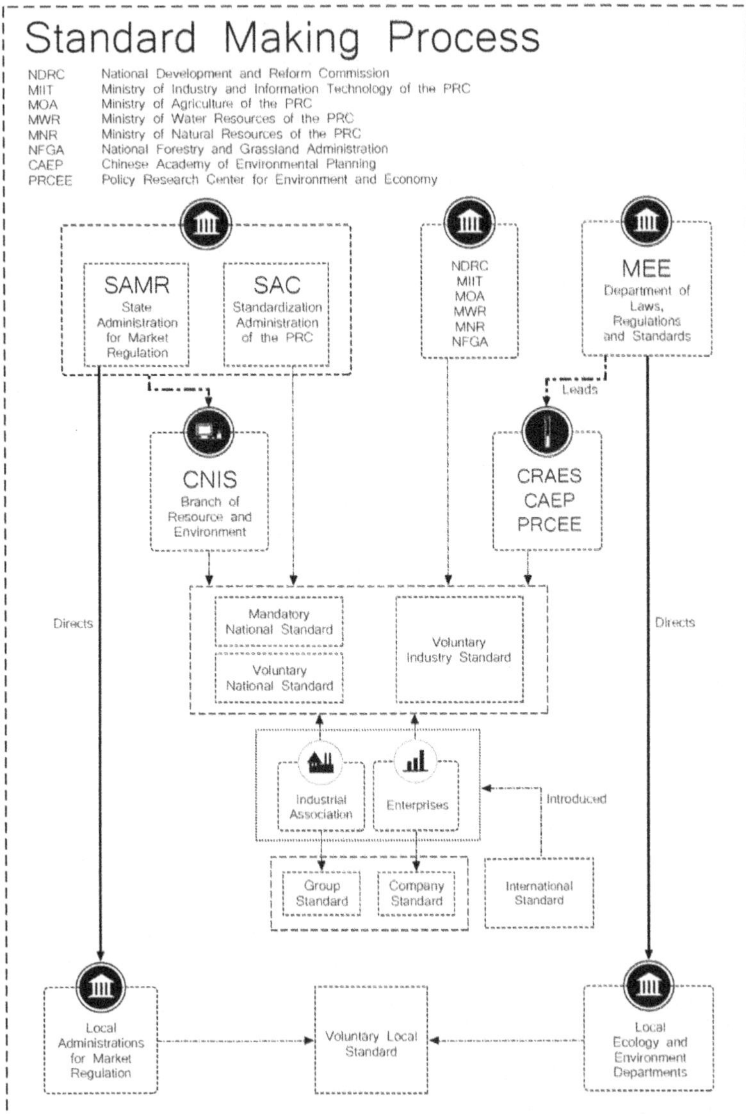

Fig. 2 Institutional actors and processes in China's standard-making

(In China's *"tiao kuai"* system, central government entities, such as the MEE and the NDRC, are mirrored at provincial and local levels.)

These local and national standards form the basis of environmental standards. Industry standards may implement those standards in industry-specific ways. Their development involves a broader range of actors. For environmental governance, a key body is MEE, particularly the Department of Laws, Regulations and Standards.[5] Other key bodies include the departments related to particular industries or industry groups, for example, the Ministry of Industry and Information Technology. CRAES and CNIS are present in the drafting of all kinds of standards, whether they are national, local or industry, voluntary or mandatory.

Among non-state actors, research institutes, universities, industry associations and enterprises can make group standards independent of government intervention. They also participate in the drafting of industry and national standards following the requests of MEE or other standard-making government bodies.

An example can better illustrate the operating mechanisms of non-state actors in environmental standard-making. Gree Electric Appliances Inc. of Zhuhai, a leading China manufacturer of household appliances such as air-conditioners and refrigerators in China, is involved in standard-setting processes. In 2005, the standardization committee of GREE released an enterprise standard called QJ/GD 92.00.001-B *Standard for the Control of Hazardous Substances in Green Products* (环保产品中有害物质控制管理标准). This standard was based on the European RoHS: Restriction of Hazardous Substances. GREE uses its private environmental label to distinguish whether the supplier follows the rules of the standard, which can be considered as a method of supervision and supply chain management.

Apart from self-deployed internal standards, GREE also participates in setting group standards developed by companies and associations together. For example, GREE and 50 other electrical appliance companies, together with the China Association for Quality Inspection, made "Construction Inspection Rules for Air Purification Systems" and "Technical Requirements for Indoor Air Quality Online Monitoring Systems."

Haier, another Chinese appliance company, also has played an important role in making industry standards, especially emission standards. In 2015, the National Center for Quality Supervision and Inspection of Indoor and In-car Air (国家室内车内环境及环保产品质量监督检验中心) and the China Indoor Environmental Monitoring Committee (中国室内环境监测工作委员会) partnered with Haier to publish an environmental industry standard called "Indoor Air Quality Assessment" based on a product of Haier.[6]

In addition, industry associations and NGOs have come to propose guidelines and recommended practices for companies to better engage in environmental impact management and in other aspects of corporate social responsibility (CSR). (See chapter "The Governance Effect of Environmental CSR Reporting in China: State and Non-state Facilitation" for more detail on China CSR.)

For example, as discussed in chapter "Extending Enforcement: How the Institute of Public and Environmental Affairs Leverages Public Information to Strengthen Environmental Governance", the Institute of Public and Environmental Affairs (IPE) has established a GCA (Green Choice Alliance) project in association with 21 other environmental organizations.[7] Greenpeace published an RPP (Responsible Purchasing Policy of Timber) that divided timber products into 4 levels and lists only the top two levels as acceptable when purchasing from suppliers. Companies that agree to join this policy should hire qualified institutes to do an annual audit, while there is another third-party audit to investigate whether the companies follow the action.[8] Greenpeace also released a DIY responsible purchasing ranking list as an encouragement.[9] (See chapter "Will China Industrial Organizations Succeed in Addressing the "Trust Deficit" Arising from Global Supply Chain Governance?" for further discussion of the relation between global non-state actors and China industry association standard setting.)

The "Real Estate Green Supply Chain," set up by the Society of Entrepreneurs and Ecology (SEE), Alxa Prefecture of Inner Mongolia, China Urban Realty Association (中城联盟), China Real Estate Chamber of Commerce (全联房地产商会), Landsea Group (朗诗绿色地产) and China Vanke Co., Ltd. (万科企业股份有限公司), began in 2016. Three NGOs and two enterprises established a five-leader action club, dedicated to developing environmental standards relating to construction sources for materials including steel, cement and lumber. The project plans to report white and blacklists according to results from third-party investigations, in hope that white list suppliers will be considered with preference when real estate companies are making a purchase.[10]

The Landscape of Environmental Governance Supervision and Certification: Institutional Actors

Figure 3, on the following page, illustrates selected institutional actors and processes of supervision and certification in environmental governance.

For standards within their jurisdiction, MEE and the Ministries of Water Resources, Natural Resources, Housing and Urban–Rural Development, Agriculture and Rural Affairs, Transport and the National Forestry and Grassland Administration, along with the corresponding local departments and bureaus, undertake environmental monitoring and supervision, often using *shiye danwei* and other non-state actors, as shown by the bottom part of the chart with MEE as an example.

Though not included in the chart, the procuratorate also plays an important supervisory role through the recent encouragement for the procuratorate to engage in environmental public interest litigation, as well as criminal prosecution. It is within the local procuratorates' mandate to conduct environmental inspection and to verify whether administrative bodies responsible for supervision and environmental law enforcement—those listed above—are effectively performing those duties. When procuratorates find cases of environmental regulation violation, they can enter into pre-litigation procedures, in which they issue pre-litigation recommendations to the administrative bodies or meet with them to discuss regulatory solutions. In cases where administrative law enforcement bodies still fail to perform after the pre-litigation

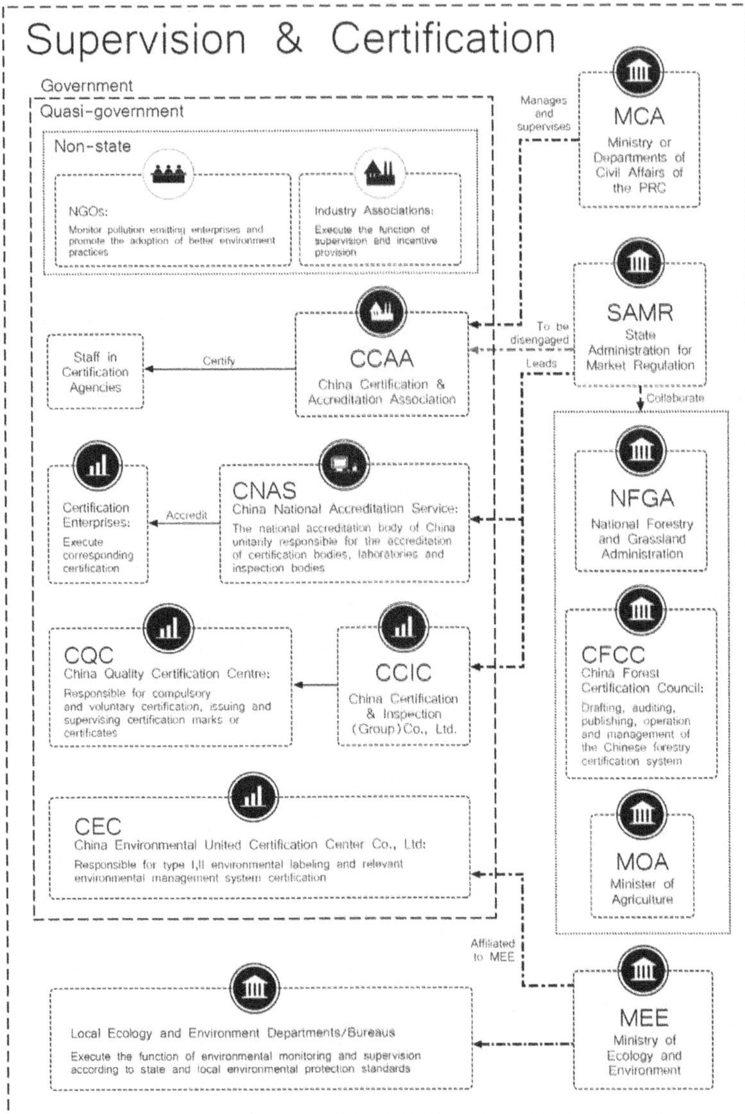

Fig. 3 Institutional actors in environmental governance supervision and certification

procedures, the procuratorate can bring public interest litigations against the administrative bodies.

Similarly, though not illustrated by this chart, Environmental Impact Assessment (EIA) as a supervision process has long been a component of China environmental law. Prior to 2018 revision of the Environmental Impact Assessment Law, preliminary environmental impact assessment in China was executed by institutions (private or public enterprises, or universities) qualified by the Appraisal Center for Environment and Engineering (ACEE), which is a *shiye danwei* affiliated with the Ministry of Ecology and Environment. The 2018 revision reformed the EIA system and "disengaged" (脱钩) EIA institutions from ecology and environment departments within the government. The amendment abolished the mechanism of administrative qualification for environmental impact assessment institutions. Firms can entrust any technical institution for the task, and if the firms themselves possess such capability, they can conduct environmental impact assessment for their own projects and write assessment reports.

The rest of the chart illustrates accreditation and certification. The China National Accreditation Service for Conformity Assessment (CNAS) accredits companies in environmental standards certification. One example is the China Building Material Test & Certification Group Co., Ltd., which is affiliated with China National Building Materials Group Co., Ltd. Though the company executes compulsory certification only on two products, i.e., safety glass and porcelain tile, it also executes voluntary certifications for products including energy-saving, low carbon and other green products. It also executes certification for Environmental Product Declarations.[11,12]

Another important body in standard certification is the China Certification & Inspection Group (CCIC); the China Quality Certification Centre (CQC) is under its control. CQC executes both compulsory and voluntary certification and undertakes the China Compulsory Certification (CCC) for 134 kinds of products to ensure compliance with basic product quality standards. For voluntary certification, CQC undertakes certification for products according to national standards and also its own standards. It also supervises the enterprises which receive certification to assure their products or production process always conform with the standards or certification.

A further important organization in environmental standard certification is the China Environmental United Certification Center Co., Ltd. (CEC). This company is affiliated with MEE and responsible for type I and type II environmental labeling.

Under MEE authorization, CEC manages the China Environmental Labeling system. The labeling foci include conformance with international treaties, recycling and reuse, regional environmental quality improvement, indoor environment improvement, human health safeguards, improvement in the efficiency of resource and energy use, and promoting energy-saving and emission-cutting, reducing greenhouse effect. Since 2006, products with China Environmental Labeling have gradually acquired advantages in government procurement projects, which provide strong motivation for manufacturing enterprises.

These certification organizations also adapt ISO (International Standards Organization) standards, for example, environmental management systems. Certification organization staff are audited by the China Certification & Accreditation Association.

Depending on their jurisdiction, other agencies are involved in environment-related certification. The Ministry of Agriculture (MOA) has a China Green Food Development Center, which is responsible for developing and promoting green food standards, auditing, quality supervision and label management. The China Forest Certificate Council is responsible for drafting, auditing, publishing, operating and managing the Chinese forestry certification system. It collaborates with the National Forestry and Grassland Administration of the PRC (NFGA) in regulating the forest certification market.[13]

In addition, industry associations may also execute supervision and certification functions, as later illustrated by issue and industry-specific sections. (See chapter "Will China Industrial Organizations Succeed in Addressing the "Trust Deficit" Arising from Global Supply Chain Governance?" for discussion of industry association roles.)

China Water Pollution Governance: The Landscape of Institutional Actors

Figure 4, on the following page, illustrates selected dynamics and actors in the landscape of water governance in China.

As discussed in further detail in chapter "Water Stewardship: Engaging Business, Civil Society and Government in Collaborative Solutions to China's Freshwater Challenges" by Michael Spencer and Xu Zhen, in China's decades of industrialization, the problem of water pollution became increasingly acute. The attached chart supplements the chapter with a focus on the role of non-state actors in China's water governance.

In 2015, the State Council issued the Action Plan for Prevention and Control of Water Pollution (水污染防治行动计划), often referred to as the "ten points for water" (水十条) since it includes ten focal areas of action. The plan includes targets (*mu biao;* 目标) for, among other items, urban domestic pollution, pipe networks, agricultural and rural pollution, ship and port pollution, as well as water pollution arising out of industrial production. The water quality of any river or lake or watershed may likely involve multiple foci and actors. The foci include intake, discharge, infrastructure and pollution-emitting entities' internal water management. In each case, government actors may include multiple standing government components. In urban areas, for example, these include standing government component bureaus (e.g., environment, water, urban development) and then, special zones, for example, economic development zones, export processing zones or industrial parks. In rural areas, there are also likely multiple sources of government authority and oversight jurisdiction. The jurisdictional issues and relationships are often addressed in documents issued by MEE, the Ministry of Housing and Urban–Rural Development (MOHURD) and the Ministry of Agriculture and Rural Affairs (MARA), as well as reports from the National People's Congress.[14]

In this context, the water landscape discussed here and in the accompanying chart focuses on industrial water pollution, which is both a significant contributor to China's overall water pollution and an important source of public indignation against pollution more generally. Industrial water pollution mainly concerns enterprises' internal controls.

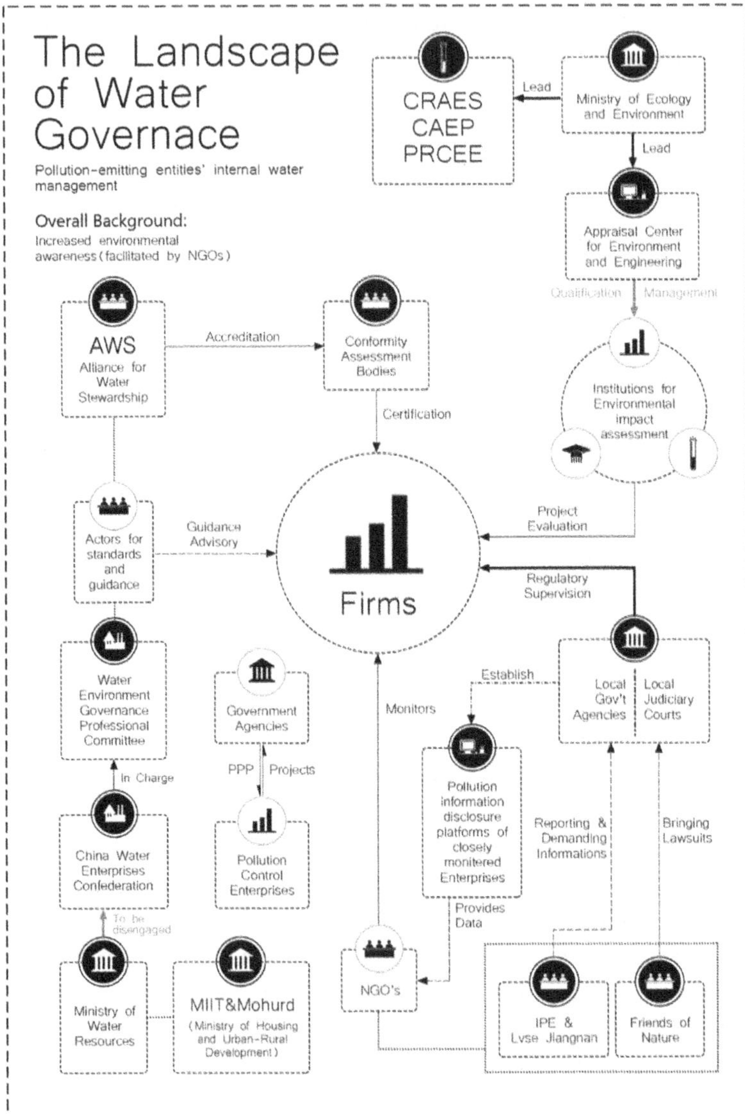

Fig. 4 Institutional actors in China's water pollution governance

The focal points of law enforcement lie in ensuring that enterprises do not discharge unprocessed wastewater stealthily, in an unregulated and illegal manner, into the natural environment, and that the wastewater discharged by these enterprises into municipalities' sewage network or wastewater processing facilities in industrial clustering zones meets relevant standards. Therefore, firms occupy the center of this chart and are the subjects of all governance actions illustrated in the chart.

Firms must be aware of government laws (including laws and regulations which directly address water pollution, but also, for example, laws/regulations that provide for procedures, such as environmental assessment or permitting), but also requirements contained in plans (such as the 2015 action plan) and further policies. In addition, as the chapter by Spencer and Xu explains, a proactive firm can also take initiative in adopting best practices recommended by NGOs, business associations and other non-state actors that provide standards and guidance to help firms reduce their environmental impact.

China Air Quality Governance: The Institutional Landscape

Figures 5 and 6 on the following pages illustrate selected institutional actors and dynamics in the landscape of air governance, with chart 5 focusing on standard-making and supervision, and chart 6 focusing on technical and academic exchanges using the electricity generation sector as an example.

Air pollution represents China's most visible environmental problem and is a source of constant public, as well as official, attention. In addition to air laws, in 2013, following the globally publicized air pollution episodes in Beijing earlier that year, the State Council issued a five year Air Pollution Prevention and Control Action Plan (大气污染防治计划), often referred to as "ten points for air" (大气十条), to map the strategy for managing air pollution. The plan covers various aspects of air governance, from targeting coal-burning boilers and power plants to encouraging the participation of the public.[15] In 2018, the State Council issued a renewed action plan called Winning the Blue Sky Defense

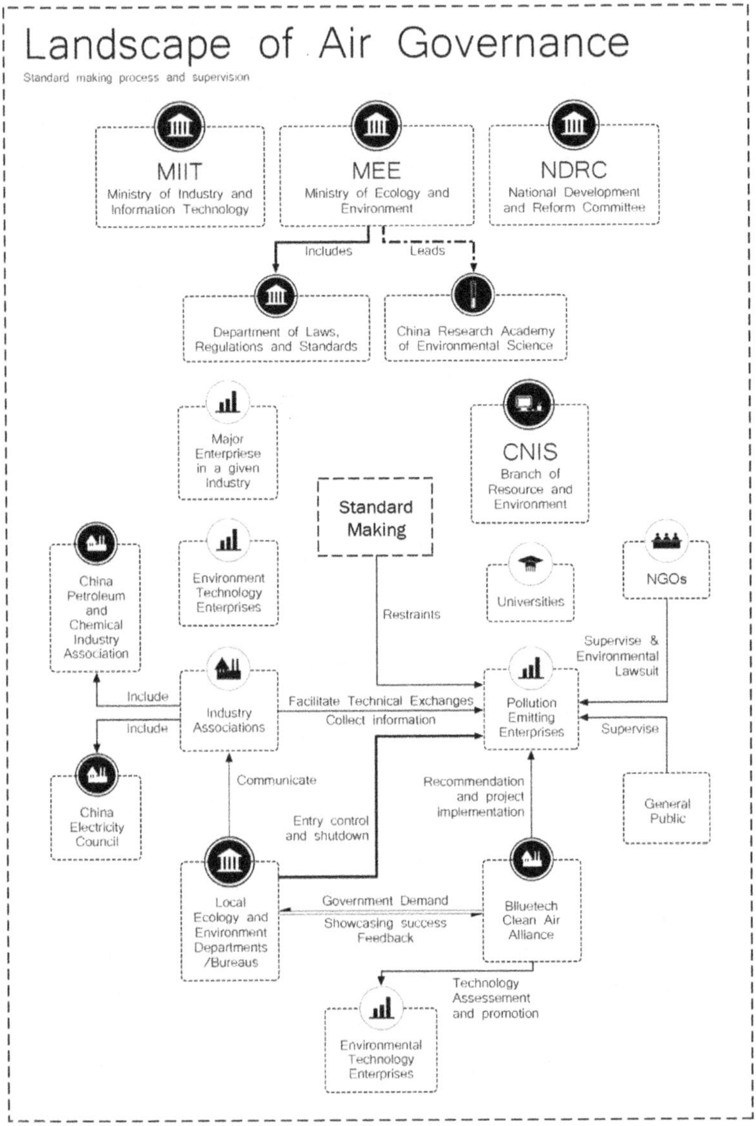

Fig. 5 Standard-making and supervision in China's air quality governance

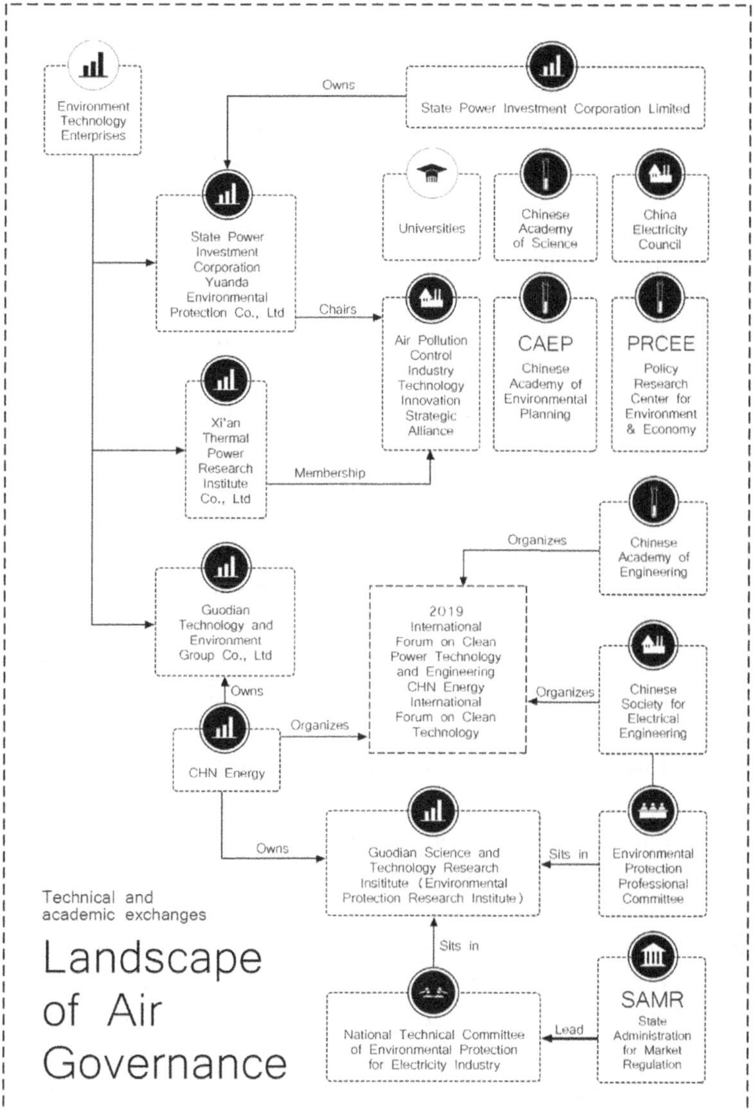

Fig. 6 Technical and academic exchanges in China's air quality governance

Battle Three Year Action Plan (打赢蓝天保卫战三年行动计划), which embraced the major principles of the 2013 plan while updating targets.[16] While the State Council plan places core responsibility on government agencies to meet targets, non-state actors (including non-state enterprises as well as state-owned enterprises) also occupy a significant space in the field.

Standard-making could be seen as the starting point of air governance, since pollutant emission standards for each industry, production process and product not only form reference points for supervision and punishment, but also guide and influence the development of green technology for a given industry. One of the most prominent actors in this field is the Chinese Research Academy of Environmental Sciences (中国环境科学研究院) and a subordinate department, the Environmental Standard Research Institute (环境标准研究所). (As noted earlier, CRAES is a *shiye danwei* affiliated with the Ministry for Ecology and Environment.) Regional environmental research institutes and universities are similarly engaged in work on air pollutant emission standard for various industries. Major enterprises in the industry as well as industry associations also participate. For example, the authors of GB 28664-2012 *Emission standard of air pollutants for steel smelt industry* (炼钢工业大气污染物排放标准) include Baoshan Iron & Steel Co., Ltd., a subsidiary of China Baowu Steel Group Corporation Limited, a central SOE (state-owned enterprise), and one of the largest iron and steel companies in the world, as well as Baosteel Engineering & Technology Group Corporation, another subsidiary of the Baowu Steel Group. The China Inorganic Salts Association, an industry association and a social group (社会团体), is among the authors of GB 31573-2015 *Emission standards of pollutants for inorganic chemical industry* (无机化学工业污染物排放标准). Standards for the methodology of measuring the concentration of a certain pollutant are generally composed by local environment monitoring stations, which are generally *shiye danwei* (事业单位) financially supported by regional government bodies. Actors involved in standard setting related to air pollution are fairly diverse, and there is no written procedure and rule for which entity to select for the composition of a given standard made public. This decision has often been within the jurisdiction of the Department of Science and Technology Standard under the MEE (now as the Department of Laws, Regulations and Standards), which may organize national environmental standard-setting processes.[17]

Actors involved in law enforcement and supervision according to the standards are also diverse. Compliance with pollutant emission standards and disclosure of emission data of "key enterprises" is under the purview of departments of ecology and environment in regional governments. Also, non-state actors including environment NGOs and citizens participate in the supervision of enterprises. Under 2014 amendment to environmental law, environmental civil society organizations that meet criteria set by the law may bring lawsuits against enterprises on behalf of environmental public interest. Citizens may participate through formal complaint processes, now often online, and less formal "*wei quan*" (e.g., web or WeChat comment). As discussed in further chapters, NGOs such as the Institute of Public and Environmental Affairs (IPE) and Lvse Jiangnan keep track of government-required emission data from enterprises on their website and mobile phone applications and take complaints from the general public. (See generally this chapter and chapter "Extending Enforcement: How the Institute of Public and Environmental Affairs Leverages Public Information to Strengthen Environmental Governance" on the work of IPE, and Part III case studies.)[18]

Air pollution control is also the focus of relevant industry associations, such as electricity generation, petroleum, chemicals, iron and steel and vehicles. (See chapter "Will China Industrial Organizations Succeed in Addressing the "Trust Deficit" Arising from Global Supply Chain Governance?" on industry associations.) They may collect information from enterprise members, assist in the initiatives of other environmental institutions and take part in promoting industry green development. For example, the China Electricity Council assisted with a study conducted by the MEE (then MEP) Environmental Impact Assessment Center in an anonymous survey of air pollutant emission compliance by thermal power plants.[19] The China Petroleum and Chemical Industry Federation makes group standard concerning "green" evaluation criteria including those for industrial parks and publishes lists of green technology.[20]

Research, development and application of relevant technology, as well as technical exchanges among firms, research institutes and other parties in the industry, are also an important air governance dynamic. Figure 6 tries to capture this dynamic with the example of electricity generation. Major electricity SOEs (state-owned enterprise or *guoyou qiye*), environmental technology companies that they own and other research institutes in this field play major roles. The five major electricity generation SOEs in China (CHN Energy, China Huaneng Group Co. Ltd, China Datang Corporation Ltd, State Power Investment Co., Ltd. and China Huadian Corporation Ltd) account for more than 43.8% of all installed capacity in China.[21] Among them, CHN Energy has the largest scale in terms of installed capacity. One of its subsidiaries, the Guodian Environmental Protection Research Institute focuses on research and development of environmental protection technology. It affiliates with the Chinese Society for Electrical Engineering in forming its professional committee for environmental protection.[22] The institute is also involved in standard setting, filling the role of the National Technical Committee of Environmental Protection for Electricity Industry, one of the many technical committees under the SAC.[23] CHN Energy also has other research institutes among its subsidiaries. The Guodian Technology & Environment Group Co., Ltd. is the major platform for environmental protection solutions for companies in the CHN Energy group; its foci include desulfurization, denitrification and water treatment.[24]

Another energy SOE, the State Power Investment Co., Ltd., is a controlling shareholder of the Yuanda Environmental Protection Co., Ltd., whose business also includes desulfurization, denitrification and smoke removal. The company chairs the Air Pollution Control Industry Technology Innovation Strategic Alliance, whose mission is to seek solutions to key technical problems, to promote adoption of advanced technology and to support policymaking concerning the industry by facilitating technical exchanges among enterprises, universities and research institutes. Members of the alliance include a number of environmental technology companies and universities, as well as the China National Institute of Standardization (CNIS) and the Chinese Research Academy of Environmental Sciences (CRAES). Another prominent member of this alliance is Xi'an Thermal Power Research Institute Co., Ltd. (TPRI), whose shareholders include the five largest Chinese energy companies.[25] One of the missions of TPRI is to reduce pollution in thermal power plants by developing clean coal electricity generation technology.[26]

Efforts for promoting exchanges among firms and research institutes include periodic global gatherings; for example, CHN Energy hosted the 2019 International Forum on Clean Power Technology and Engineering and the CHN Energy International Forum on Clean Technology along with the Chinese Academy of Engineering and the Chinese Society for Electrical Engineering.[27] China Electricity Council's Energy Saving and Environmental Protection sub-council also organizes conferences for exchanges and dialogues, with focus on addressing difficulties encountered by member enterprises in technical upgrades.[28]

In the broader inter-industry context of technical exchanges and technology application, one example is the Bluetech Clean Air Alliance, shown in Figure 5 because of its relationship with government bodies. The alliance is a non-profit social group (*shehui tuanti*) dedicated to environmental technology introduction, technical transfer, assessment, demonstration and promotion. Members of the alliance include various environmental technology companies, universities, research institutes and international NGOs.[29]

The Landscape of Environmental Governance in the Air-Conditioning and Refrigeration Industry: Institutional Actors

In chapter "Non-industry and Nonstate Actors Contribution in the Standard Drafting Process: Examples from the Development of China Room Air Conditioner Standards", Richard ("Tad") Ferris and Sun Xiaopu focus on the role of non-state actors in the process of government environment/sustainability-related standard setting using the air-conditioning governance context. In 2020, these efforts include development of China air-conditioning standards following the 2016 amendment to the Montreal Protocol and heightened efforts to raise the bar on energy efficiency standards around the time of publication of the Action Plan for Green, High Efficiency Cooling and Refrigeration in June 2019.[30] Figure 7 uses this context to provide an example of industry-specific environmental governance.

The National Development and Reform Committee (NDRC) and Ministry of Ecology and Environment (MEE) are the core state actors involved in promoting energy efficiency and paving the way for new refrigerant adoption. Among non-state actors, enterprises, research institutes and universities are active in driving for technological innovations and their industry applications. They form industry knowledge and information clusters, from which standard-making efforts also benefit.

Industry associations also contribute to the industry knowledge and information clusters. The China Refrigeration and Air-conditioning Industry Association (CRAA) is a business association (also a social group, *shehui tuanti*) comprised of manufacturers within the industry and relevant research institutes and universities. CRAA seeks to build bridges between government and member organizations and facilitate cooperation in improvement of product competitiveness and industry advancement. For example, the 2018 China Refrigeration and Air-conditioning Industry Information Conference, organized by CRAA, included a forum dedicated to new refrigerant development following the Montreal Protocol and to progress in energy efficiency improvement for products using the new refrigerants. Speakers include three university professors, one researcher from the China National Institute of Standardization, and managers and engineers from relevant enterprises. Their reports presented characteristics of various new refrigerant types, including R32、R1234yf, R1234ze(E), R513A and HFOs, their refrigerating capacity and efficiency, their compatibility with extensively-used equipment, testing results obtained by major producers, and implications of energy efficiency improvement and new refrigerant adoption for the standard-making for this industry.[31]

The CRAA also cooperates with governmental agencies and international organizations in various actions aimed at new refrigerant adoption. Within the MEE, there is the MEE Foreign Economic Cooperation Office (FECO), and the cooperation between FECO and CRAA includes formulating plans for phasing out HCFCs and HFCs, encouraging enterprises' production line conversion, assisting

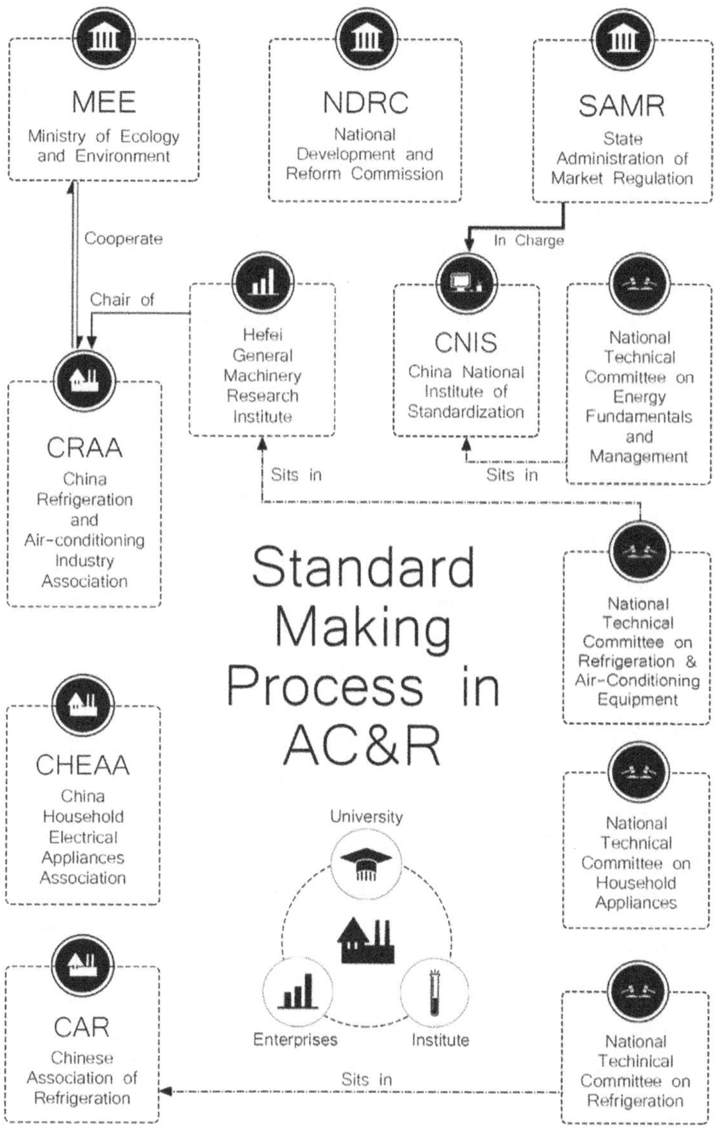

Fig. 7 Institutional actors in environmental governance in the air-conditioning and refrigeration industry

small and medium-sized enterprises (SMEs) in obtaining technological and funding support, and organizing industry activities.[32]

The Chinese Association of Refrigeration (CAR) focuses on academic exchanges. It organizes a biennial academic conference, financially supported by the industry. The CAR edits and publishes the Journal of Refrigeration and further professional literature and technical materials. It also cooperates with the CRAA in holding the China Refrigeration Exhibition.[33]

Another industry association that is involved in new refrigerant adoption and energy efficiency improvement is the China Household Electrical Appliances Association (CHEAA), whose members include manufacturers of home air-conditioners and producers of other appliances that use refrigerants. This association also organizes its own conferences and works with FECO in production line conversions and relevant industry capacity building.[34]

In addition to industry associations, the Hefei General Machinery Institute (GMRI), belonging directly to China National Machinery Industry Corporation Ltd (SINOMACH), a central SOE, engages in research, design, manufacturing and engineering of refrigeration and air-conditioning equipment.[35] Experts from the GMRI participate in research for new refrigerants, review and acceptance of converted refrigeration production line, and in-house training for employees in companies that are converting to new refrigerants.[36] GMRI also has subsidiaries that focus on development, manufacturing and construction of energy-saving products and integrated equipment sets. Hefei General Machinery Product Institute Co., Ltd. carries out testing on refrigeration and air-conditioning products and certifies energy-saving products, while the Hefei General Environmental Control Technology Co., Ltd. designs and supplies those testing devices. Hefei Haoke Chemical Equipment Engineering Company carries out technology and product upgrades.[37]

As non-state actors with strong presence in academic and technical exchanges in the air-conditioning and refrigeration industry, CRAA, CAR and GMRI are equally present in standard-making initiatives in the field. CAR and GMRI host secretariats of two technical committees under the Standardization Administration of China—the National Technical Committee on refrigeration and the National Technical Committee on Refrigeration and Air-conditioning Equipment, respectively.

These non-state actors play continued roles in the development of standards related to energy efficiency and new refrigerant development. The mandatory national standard GB 21455-2019 *Minimum allowable values of the energy efficiency and energy efficiency grades for room air conditioners* is seen as a seminal effort to raise the bar on air-conditioners' energy efficiency and to phase out less efficient products. It was proposed by the SAC and the Department of Resource Conservation and Environmental Protection under the NDRC. Its drafters include major air-conditioner and air-conditioning equipment producers, three universities and the Hefei General Machinery Product Certification Co., Ltd., a subsidiary of GMRI. In new refrigerant development, for example, R290 is a promising refrigerant for air-conditioners, but it is more flammable.[38] Its adoption requires corresponding updates of safety standards. In 2016, QB/T 4835-2015 *Technical requirements for installation service and transportation of room air conditioner employed flammable refrigerants* came into effect. This standard was proposed by the China National Light Industry Council, drafted by the CHEAA, Huazhong University of Science and Technology, five major air-conditioner manufactures and a refrigerant manufacturing company, and is managed by the National Technical Committee on Household Appliances.[39] In 2018, GB/T 9237—2017 *Refrigerating systems and heat pumps – safety and environmental requirements* began to be implemented. This standard, modeled after ISO 5149:2014 with some modifications, provides instructions for usage of flammable refrigerants including R290 and R32. It was proposed by the China Machinery Industry Federation and its drafters include the GMRI, CRAA, Tianjin University and various refrigeration enterprises. This standard is managed by National Technical Committee on Refrigeration & Air-Conditioning Equipment under GMRI.[40]

The Landscape of Environmental Governance in the Textile Industry

Figures 8 and 9 on the following pages supplement the chapter "Will China Industrial Organizations Succeed in Addressing the "Trust Deficit" Arising from Global Supply Chain Governance?" by Juan Zhang on China industry or business associations (shehui tuanti) with a corresponding focus on the textile industry. Figure 8 focuses on standard-making while Figure 9 presents examples of further environmental governance initiatives within the industry. The textile industry, including dying and printing, is one of the most polluting industries in China. Pollution from this industry has caused substantial water quality degradation in rivers and lakes, with episodes in locale including Lake Tai in the Yangtze River delta gaining global attention.

In addition to domestic pressure, textile companies in China face requests from importers and international brands for compliance with certain social responsibility standards, as international brands become increasingly dedicated to supply chain management. (See Lydia Price's chapter "The Governance Effect of Environmental CSR Reporting in China: State and Non-state Facilitation" on China enterprise Corporate Social Responsibility and Juan Zhang's chapter "Will China Industrial Organizations Succeed in Addressing the "Trust Deficit" Arising from Global Supply Chain Governance?" on China industry association voluntary standards.)

In 2012, in response to the exacerbation of environmental problems caused by the textile industry, the Ministry of Ecology and Environment (then the Ministry of Environmental Protection) issued a new standard of industrial wastewater discharge (GB 4287-2012 *Discharge Standards of Water Pollutants for Dyeing and Finishing of Textile Industry*).[41] The Ministry of Environmental Protection Department of Science and Technology Standards (now the Ministry of Ecology and Environment's Department of Laws, Regulations and Standards) organized the drafting of the aforementioned water discharge standard.

The Chinese Research Academy of Environmental Sciences (CRAES), a *shi ye danwei* affiliated with MEE, plays a key role in technical support to MEE, for example working on standards including HJ/T185-2006

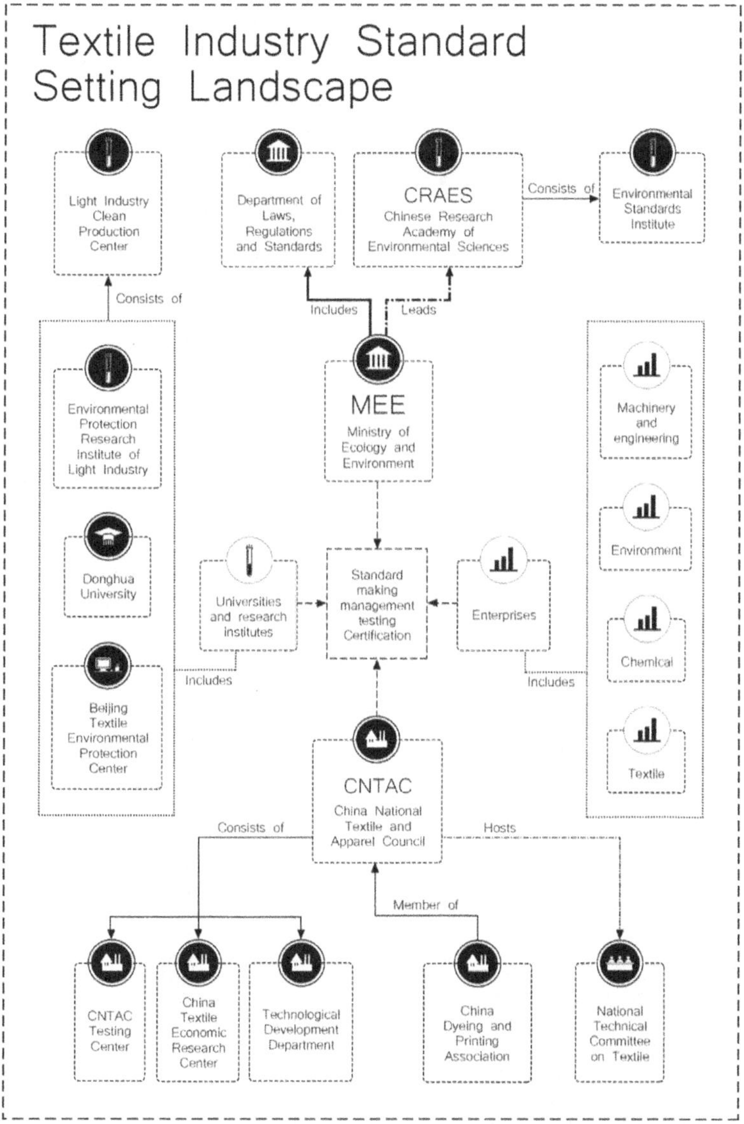

Fig. 8 Institutional actors in standard-making in the textile industry

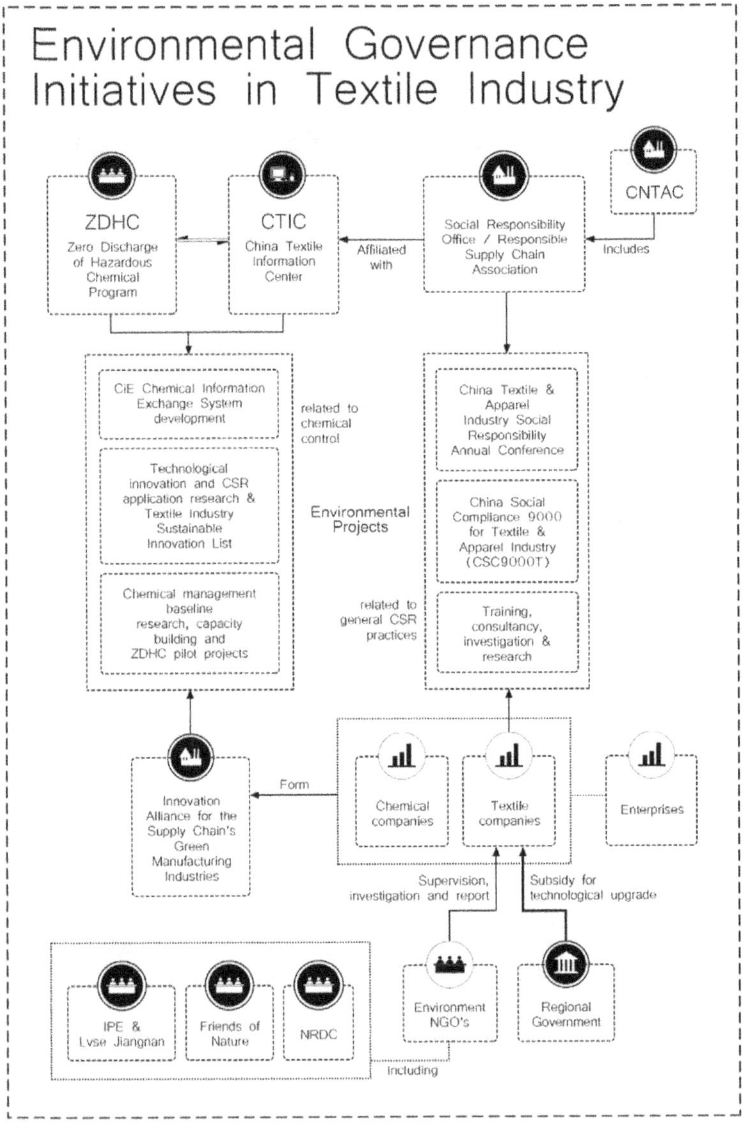

Fig. 9 Further environmental governance initiatives in the textile industry

Cleaner Production Standard – Textile Industry (Dyeing and Finishing of Cotton) and HJ990-2018 *Technical Guidelines of Accounting Method for Pollution Source Intensity – Textile and Dyeing Industry.*[42] MEE is also responsible for investigation and supervision of the enterprises' compliance with environment regulations.

In addition, textile standard setting involves further industry associations, enterprises, university and research institutes. For example:

- the drafters of the aforementioned discharge standards of water pollutants included the China Textile Economic Research Center under the China National Textile and Apparel Council (CNTAC), Donghua University and Furun Holding Group Co., Ltd.
- the pollution source intensity standard (HJ990-2018) drafters include the China Light Industry Clean Production Center, which is under the Environmental Protection Research Institute of Light Industry, a Beijing-based *shiye danwei* (public service research institute).
- in September 2018, the MEE published a draft of a revised standard *Technical Specification for Dyeing and Finishing Wastewater Treatment of Textile Industry* to solicit opinions from institutions and enterprises. While the revision was organized by the MEE, principal drafters include the China Association of Environmental Protection Industry, Donghua University, China Dyeing and Printing Association (CDPA), Tsinghua University, and engineering and environmental technology companies. Institutions whose opinions were solicited included city planning and engineering companies, textile, dyeing and printing companies and environment technology companies.[43]

The organizations that undertake the making of specific textile industry-related standards depend on relevance of their expertise as well as decisions by institutions that organize the standard-making procedures. In general, the MEE organizes the drafting of emission and environmental quality standards, while CNTAC (the textile industry association, see chapter "Will China Industrial Organizations Succeed in Addressing the "Trust Deficit" Arising from Global Supply Chain Governance?"), other textile industry associations and textile research institutes such as the China Textile Academy are predominant in standard-making relating to textile products and manufacturing processes. These two areas do sometimes overlap, as CNTAC proposed an industry standard for reusing water for textile dyeing and finishing (FZ/T01107-2011) which was then drafted by Donghua University and CDPA.[44]

In addition to government, enterprises and business associations, environmental "NGOs" (*minban fei qiye*) (see this chapter), such as IPE (see chapter "Extending Enforcement: How the Institute of Public and Environmental Affairs Leverages Public Information to Strengthen Environmental Governance") and environmental NGOs qualified to bring public interest lawsuits play oversight roles. By 2018, for example, IPE had published five reports on the textile industry. The reports led to wider dialogue among CDPA, CNTAC, dyeing and printing enterprises and environmental NGOs including IPE, Natural Resources Defense Council (NRDC) (a US origin NGO) and Envirofriends.[45]

As discussed in chapter "Will China Industrial Organizations Succeed in Addressing the "Trust Deficit" Arising from Global Supply Chain Governance?", to improve textile companies' awareness and management of environment (and further social) responsibility, CNTAC has issued a textile industry-specific guideline (CSC9000T) and holds training sessions, conferences and carries out research and investigations in textile enterprises to promote adoption of the measures suggested by CSC9000T.

CNTAC, in turn, forms and/or participates in further relationships. For example, CNTAC formed a strategic relationship with an NGO, the Zero Discharge of Hazardous Chemical program (ZDHC) in 2016 to promote ZDHC's Chemical Stewardship Initiative 2020 and its 2020 Target for Hazardous Chemical Zero Emission Project.[46] This initiative is carried out with the China Textile Information Center (CTIC) and the Innovation Alliance for the Supply Chain's Green Manufacturing Industries (TexSMC).[47] The CTIC is a *shiye danwei* built by NDRC and managed by CNTAC that conduct consulting, research and other information-related service for the textile industry.[48] TexSMC, founded in the 2015 China Textile & Apparel Industry Social Responsibility Annual Conference, was initiated by the CNTAC office for social responsibility and consists of 11 textile chemical companies and 14 textile companies. In 2018, members of TexSMC declared the initiation of industry self-governance initiative to compile a list of controlled chemical substances in the supply chain and their discharge standards, to come out with management guidelines of those chemicals and to participate in standardization efforts. The alliance also looks to support the utilization of the textile chemical environmental impact information exchange system (also known as CiE) to disclose information about chemical production, usage and discharge in the supply chains.[49]

Notes

1. See https://www.samr.gov.cn/jg/, https://www.sac.gov.cn/zzjg/jgzz/#, https://www.cnis.ac.cn/bygk/byjj/, https://www.mee.gov.cn/zjhb/.
2. See https://www.cnca.gov.cn/, ccaa.org.cn, cnas.org.cn/index.shtml, https://www.ccai.cc/, https://www.ccic.com/, https://en.mepcec.com/.
3. See https://www.caep.org.cn/, https://www.prcee.org/, https://www.craes.cn/, https://iue.cass.cn/.
4. Standard information is available on the National Public Service Platform for Standards Information: https://std.samr.gov.cn/.
5. See https://www.mee.gov.cn/zjhb/bjg/fgs/ for the functions of this department, which include the management of national environmental standards. Accordingly, there are sections of Law, Regulations and Standards within local ecology and environment bureaus in charge of local standards and regulation. See, for example, the Shanghai EEB: https://sthj.sh.gov.cn/sh/list_new.jsp?channelid=2002; https://sthj.sh.gov.cn/sh/list_new.jsp?channelid=2014.
6. "国家机构联合海尔空调推出中国室内空气两大标准 | 市场动态 | 新闻动态 | 关于海尔 | 海尔集团."
7. "揭秘中国水污染地图和绿色选择联盟–环保–人民网."
8. See https://www.greenpeace.org.cn/china/Global/china/_planet-2/report/2007/11/responsible-purchase.pdf.
9. "'家居建材零售连锁企业负责任采购木材'的第一份成绩单 |."
10. "绿色供应链 - 阿拉善SEE基金会."
11. This is also the Type III declaration under ISO 14,025. See: https://www.iso.org/standard/38131.html.
12. See https://www.ctc.ac.cn/col/col70/index.html.
13. See https://www.greenfood.agri.cn/ and https://www.cfcc.org.cn/.

14. See the Three-Year Action Plan for Quality and Effectiveness Improvement in Urban Sewage Processing (城镇污水处理提质增效三年行动方案) issued by MOHURD, MEE and NDRC, the Action Plan for the Battle of Pollution Governance in Agriculture and Rural Areas (农业农村污染治理攻坚行动方案) issued by the MEE and the MARA, and the Report by the NPC Standing Committee Law Enforcement Inspection Group on the Inspection on Implementation of the Water Pollution Prevention and Control Law of the People's Republic of China (全国人大常委会执法检查组关于检查《中华人民共和国水污染防治法》实施情况的报告).
15. "国务院关于印发大气污染防治行动计划的通知," accessed January 3, 2020, https://www.gov.cn/zwgk/2013-09/12/content_2486773.htm.
16. "国务院关于印发打赢蓝天保卫战三年行动计划的通知 (国发〔2018〕22号) _政府信息公开专栏," accessed January 3, 2020, https://www.gov.cn/zhengce/content/2018-07/03/content_5303158.htm.
17. See https://www.mee.gov.cn/zjhb/bjg/fgs/.
18. "环境日启动, 生态环境部首次聘请10位 '特邀观察员' - 国内 - 新京报网," accessed August 15, 2019, https://www.bjnews.com.cn/news/2019/06/02/586491.html.
19. "中电联关于开展火电厂大气污染物达标排放调查工作的通知 (中电联行环函[2017]115号)," accessed August 15, 2019, https://huanzi.cec.org.cn/huanbaobu/2017-09-19/173248.html.
20. "8项任务推进石化业绿色发展 - 中化新网," accessed January 3, 2020, https://www.ccin.com.cn/detail/55bd4879c5a6f89154b51756b1c4255c/news.
21. "集团概况," accessed January 5, 2020, https://www.ceic.com/gjnyjtww/chnjtjs/chnjtjs.shtml; "我们的企业," accessed January 5, 2020, https://www.chng.com.cn/n31529/n31551/index.html; "中国大唐集团有限公司," accessed January 5, 2020, https://www.china-cdt.com/dtwz/showfirst/index_gydt&s=gydt_gsjj.html; "中国华电集团有限公司-公司概况," accessed January 5, 2020, https://www.chd.com.cn/webfront/webpage/web/contentList/channelId/41d2af3e5ba34d4cb5184b1579644376/pageNo/1; "国家电力投资集团有限公司," accessed January 5, 2020, https://www.spic.com.cn/jtgk_dhzy/jtjs/; "中电联发布《中国电力行业年度发展报告2019》," accessed January 5, 2020, https://www.cec.org.cn/yaowenkuaidi/2019-06-14/191782.html.
22. See https://www.csee.org.cn/portal/zzjgzywyh/.
23. See https://www.cec.org.cn/huanbao/jienenghbfenhui/guanyuwomen/zhuanjiafengcai/huanbao/2012-07-31/88210.html.

24. See https://www.cec.org.cn/huanbao/jienenghbfenhui/guanyuwomen/huiyuanfengcai/huanbao/2012-08-21/89512.html, https://www.ceic.com/gjnyjtww/chnzzjg/chnzzjg.shtml.
25. See https://www.zdydep.com/group/introduction/, https://www.chinaapca.com/.
26. See https://www.tpri.com.cn/.
27. See https://ceforum.ceic.com/.
28. See https://huanzi.cec.org.cn/dongtai/.
29. See https://www.cleanairchina.org/?menuId=6949.
30. "《绿色高效制冷行动方案》印发_滚动新闻_中国政府网," accessed January 3, 2020, https://www.gov.cn/xinwen/2019-06/19/content_5401443.htm.
31. See https://www.chinacraa.org/about.aspx, https://www.chinacraa.org/craa_news_show.aspx?id=3679.
32. "中国工商制冷空调行业中小企业申报第二阶段HCFCs淘汰改造项目培训会在京召开 - 制冷剂替代 - 中国制冷与空调网," accessed January 3, 2020, https://www.chinarefac.com/news_show.aspx?id=9037; "中国工商制冷空调行业HCFCs淘汰及HFCs削减战略研讨会在京召开 - 国内资讯 - 中国制冷空调工业协会," accessed January 3, 2020, https://www.chinacraa.org/news_show.aspx?id=2301; "'R22模块热泵机组采用R32替代技术改造项目'通过现场验收 - 协会动态 - 中国制冷空调工业协会," accessed October 20, 2019, https://www.chinacraa.org/craa_news_show.aspx?id=3997; "环境保护部环境保护对外合作中心成功验收R32替代项目 - 国内资讯 - 中国制冷与空调网," accessed January 3, 2020, https://www.chinarefac.com/news_show.aspx?id=307.
33. See https://www.car.org.cn/index.php?s=/lists_1.html#enyu1, https://www.car2019.cn/, https://www.car.org.cn/index.php?s=/lists_63.html.
34. "新闻动态 — 生态环境部对外合作与交流中心," accessed January 3, 2020, https://www.mepfeco.org.cn/dtxx/xwdt/201703/t20170329_565833.html; "中国家电应成为环保技术的全球引领者 - 空调设备及系统运行节能国家重点实验室," accessed January 3, 2020, https://www.acskl.com/nd.jsp?id=25.
35. See https://www.hgmri.com/index.asp.
36. "'R22模块热泵机组采用R32替代技术改造项目'通过现场验收 - 协会动态 - 中国制冷空调工业协会"; "天加HCFC淘汰改造项目通过国家环保部验收-公司新闻-天加环境科技有限公司," accessed January 3, 2020, https://www.ticachina.com/index.php?m=Article&a=show&id=231.
37. See https://www.hgmri.com/index.asp.

38. "R32制冷剂与R290制冷剂物理特性," accessed January 3, 2020, https://www.crrchina.com/news/r32290.html.
39. See https://std.samr.gov.cn/hb/search/stdHBDetailed?id=8B1827F22194BB19E05397BE0A0AB44A.
40. See https://std.samr.gov.cn/gb/search/gbDetailed?iftd=71F772D8298BD3A7E05397BE0A0AB82A.
41. "《纺织染整工业废水治理工程技术规范（征求意见稿）》编制说明," accessed January 31, 2019, https://www.mee.gov.cn/gkml/sthjbgw/stbgth/201809/W020180917349315922106.pdf.
42. See https://kjs.mee.gov.cn/hjbhbz/bzwb/other/qjscbz/200610/t20061001_78200.htm and https://kjs.mee.gov.cn/hjbhbz/bzwb/wrfzjszc/201901/W020190110594402161385.pdf.
43. See https://www.mee.gov.cn/gkml/sthjbgw/stbgth/201809/W020180917349315922106.pdf.
44. See https://std.samr.gov.cn/hb/search/stdHBDetailed?id=8B1827F1B2E5BB19E05397BE0A0AB44A.
45. "纺织4期-谁来守住污水处理的责任底线.Pdf," accessed January 31, 2019, https://wwwoa.ipe.org.cn//Upload/IPE-Reports/%E7%BA%BA%E7%BB%874%E6%9C%9F-%E8%B0%81%E6%9D%A5%E5%AE%88%E4%BD%8F%E6%B1%A1%E6%B0%B4%E5%A4%84%E7%90%86%E7%9A%84%E8%B4%A3%E4%BB%BB%E5%BA%95%E7%BA%BF.pdf.
46. "'纺织之光'纺织行业化学品管控重点成果现场推广活动上虞举行," accessed January 31, 2019, https://www.csc9000.org.cn/news/responsibility/2016-06-15/100.html.
47. "2016–2017 Annual Report on Social Responsibility of Chinese Textile and Apparel Industry," accessed January 31, 2019, https://www.csc9000.org.cn/d/file/download/CSC9000T/2018-04-11/fa4b6a31772610e9311ec14021b1053c.pdf.
48. In the organization structures of CNTAC and CTIC, CTIC is a member organization of CNTAC, but the office for social responsibility is under both CNTAC and CTIC. In fact, CTIC acts as the representative of CNTAC and manages the office for social responsibility under a structure known as "joint office" (合署办公).
49. "纺织供应链企业开展环保自治行动-新华网," accessed January 31, 2019, https://www.xinhuanet.com/energy/2018-04/18/c_1122699246.htm.

The Institutional Landscape: Some Key Organizational Types

The Governance Effect of Environmental CSR Reporting in China: State and Non-State Facilitation

Lydia J. Price

Introduction

Corporate Social Responsibility (CSR)—a term that originated in the West—is now a global concept. In its late twentieth-century western origins, the concept involved voluntary efforts by business and was sometimes referred to as business self-regulation. Organizational CSR practices and programs have been shaped over time by various external forces including market demands, community pressures, and government guidelines. In its twenty-first-century embrace by China, the concept is strongly shaped by government direction of corporate activity. In its most direct form, this government shaping manifests as environmental improvement targets embedded in Five Year Plans (FYP); in a more indirect form, it appears in requirements for CSR reporting. In this chapter, we will see the interface of those two shaping forces, and

L. J. Price (✉)
China Europe International Business School, Shanghai, China
e-mail: plydia@ceibs.edu

D. Guttman et al. (eds.), *Non-state Actors in China and Global Environmental Governance*, Governing China in the 21st Century,
https://doi.org/10.1007/978-981-33-6594-0_4

we will look into the landscape of non-state actors that play an important role in improving the implementation of these government-directed shaping forces.

Non-state actors, as provided by the landscape in chapter "The State, Nonstate Actors, and China's Environmental Performance: Setting the Stage", play a substantial role in CSR rule development and implementation. In China, state-owned enterprises, notably energy companies, are a core source of environmental challenges. At the same time, however, non-state actors play a core role in developing and implementing the rules. China's CSR reporting requirements are implemented by the China Securities Regulatory Commission (CSRC),[1] which is legally a *shi ye danwei* (as discussed in chapter "Introduction"). The CSRC, in turn, requires a subset of companies listed on the Shenzhen (SZSE) and Shanghai (SSE) stock exchanges, both *shi ye danwei* (see chapter "Introduction") to issue an annual CSR report. As of 2008, companies included in the SZSE "100 Index" and the SSE "Corporate Governance Sector" list are subject to *Mandatory CSR reporting (MCSR)*, as are the SSE-listed firms from the finance industry or with shares listed overseas.[2] Chinese companies generally tend to report on their CSR efforts to meet the government's economic and social priorities such as quality control, innovation, labor relations, and poverty alleviation (Shanghai University of Finance and Economics and China Europe International Business School 2018; UNEPFI et al. 2019), but if they operate in heavily polluting industries or they have been flagged as a potential or actual environmental offender, they must report on environmental performance as well. Companies from the metallurgy, extraction, and construction material industries fall under this more focused *Mandatory Environmental-CSR Reporting (MECSR)* requirement, but for many other listed companies, Environmental CSR (ECSR) reporting remains entirely voluntary.

Though *MESCR* has boosted transparency about China's most polluting industries and companies, the overall effect on environmental governance remains limited. One problem is that China's capital markets make very few concrete demands about what is acceptable in reporting content or style. Cultural values such as filial piety (respecting the hierarchical ordering of Confucian society—in this case, elevating the state's priorities above those of private entities) are thought to compound the problem by inhibiting firms from taking voluntary action to improve their environmental performance or environmental reporting beyond the state's minimal demands (Hofman et al. 2017). Consequently, Chinese CSR reports are often considered to be of poor quality (Hu et al. 2018) in that they contain little information that can be used to direct market penalties and rewards according to environmental performance. As such, they do little to foster self-governing behavior.

Fortunately, emerging evidence shows that non-state actors have a beneficial effect in boosting China's ECSR reporting quality in ways that have the potential to enhance governance. In addition to the core roles of the China Securities Regulatory Commission and Shanghai and Shenzhen stock exchanges, organizations that specialize in Environmental, Social, and Governance (ESG) standard-setting and reporting such as ISO (International Standards Organization) and GRI (Global Reporting Initiative)—both international non-state non-profit organizations—help companies to identify appropriate environmental metrics to report and also urge them to track data that facilitate temporal and cross-sectional benchmarking. China's Academy of Social Sciences (CASS)—also *shiye danwei* as discussed in chapter "Introduction"—has issued guidelines similar to GRI, and analysts report that growing numbers of Chinese listed companies adopt GRI, ISO, or CASS protocols each year as the basis for their CSR reporting (WTO and Golden Bee 2017). The work of these organizations, as we will discuss, is supplemented by a growing array of reports from global and domestic non-state advocacy organizations and research institutes.

A second-order effect of teaching companies to collect and disclose high-quality environmental data is a growing awareness of the links between business conduct and environmental impact. Astute firms will build on this awareness to launch internal improvement initiatives that ultimately might prove to be transformative. Non-state actors such as sustainability consultants and conference organizers, trade associations, and other advisory bodies facilitate a company's shift from a box-ticking style of environmental reporting to one that is more likely to prompt internal transformation.

But even as standardized protocols and consultants help to raise the quality and transformative power of Chinese ECSR reporting, the overall impact on environmental governance remains small because of the limited number of firms subject to the most stringent reporting mandates. Though heavy industry bears disproportionate responsibility for emissions of many pollutants, the practices of nearly every industrial sector can be harmful to the environment. All energy-intensive businesses, for example, potentially contribute to global warming. By limiting *MECSR* to heavy industry and explicit environmental offenders, the CSRC guidelines do little to instill a sense of environmental accountability on companies from other sectors. Fortunately, other non-state actors—specifically, overseas buyers and investors who face demands in their own markets for environmental transparency and performance improvements—increasingly pressure and support their Chinese suppliers and investees to improve on environmental governance and to publicize their efforts in CSR reports. Hence, non-state actors who engage with Chinese companies in business dealings can contribute positively to China's collective governance system by encouraging their partners to engage in high-quality Voluntary ECSR reporting.

To gain perspective on how non-state actors currently influence China's ECSR reporting, this chapter includes an analysis of the CSR reports of 718 Chinese listed companies for the year 2016. Results show that companies adopting a standardized protocol and/or working with foreign buyers and investors are more likely than others to report hard data about their efforts to improve on environmental problems, although in absolute terms that pool of companies remains small. Protocols helped the companies to better track and report on *past* environmental performance but were not effective at urging them to set future targets for improvement. Subsequent investigation suggests that companies are referencing the standardized protocols to learn about good reporting practices while not yet attempting to issue reports that fully comply with best standards. Furthermore, though overseas buyers and investors exert a positive influence on Chinese ECSR reporting quality, that effect is much weaker than the influence of the standardized protocols.

In the following sections, we present the results of the investigation. To set the stage, we first present a conceptual framework for the possibility of ECSR reporting to motivate company-level self-governance, as well as a discussion of how ECSR reporting is evolving in practice. After detailing the results of our studies, we conclude with a consideration of how both state and non-state actors can strengthen the potential governance impact of ECSR reporting in the future.

Two Paths from ECSR Reporting to Self-Governance

ECSR reporting in principle facilitates a company's shift toward self-governance by two key means, specifically the mandate to *disclose* information to the public and the effect of this mandate on company *transformations* (Eccles and Serafeim 2015). The first mechanism of mandated disclosure has the goal of reducing information asymmetries that distort the allocation of market rewards and punishments based on a company's environmental track record. Reporting is designed for the information needs of external stakeholders in this first case, empowering investors, buyers, and partners to "vote with their feet and wallets" according to what they consider to be good business behavior. It also opens the way for government regulators to impose or lift sanctions. As a mechanism to drive self-governance, disclosure depends on the company's belief or observation that these external stakeholders will fairly reward/punish their good/bad behavior. It also demands that stakeholders perceive the data they receive to be accurate and relevant to their decisions. Under those conditions, companies will adjust in ways that elicit market rewards and avoid penalties. Too often, however, stakeholders find publically available information to be too complex, incomplete, or untrustworthy to drive their decisions. The power of disclosure as a governance tool is thus blunted when data quality is poor.

Reporting's second function—internal company transformation—is less dependent on information quality to provoke initial action. It is a natural outcome of organizational learning (Zadek 2004), which can start slowly from weak or limited insights and improve over time. The phenomenon manifests when the process of compiling, interpreting, and releasing data yields insights that prompt the company to engage in more extensive or targeted internal environmental performance improvements. For example, a company that studies others' ECSR reports in order to prepare its own report might discover that resource conservation leads to cost control, after which the novice firm will launch a search for resource conservation opportunities of its own. Alternately, a company's leaders might discover through the process of internal data collection that employees have ideas about innovations or product improvements that would solve environmental problems or attract environmentally sensitive customers, prompting them to allocate more resources to environmental R&D. No matter the origins of a company's transformation, it will ultimately be greatest if newfound understandings can be leveraged to lower risks, increase operational efficiencies and/or boost sales. To realize these benefits, the company must work to embed environmental management in business units other than environmental reporting, which too often is delegated only to the Public Relations and Human Relations departments. A growing body of research documents the financial gains to be reaped by companies that link environmental sustainability with financial and strategic priorities, and consequently embed the work in all departments including Strategy, Marketing & Sales, Research and Operations (Kotsantonis and Serafeim 2016).

ECSR Reporting in Practice

Financial Reporting

Stock exchanges tend to establish reporting guidelines with the intent of reducing investor risks rather than boosting investee transformations. Disclosure helps investors to negatively screen out companies facing disproportionate risks of falling revenues or rising costs due to tariffs, regulations, cleanups, boycotts, demand shifts, staff turnover, raw material upgrades, or any other profit-suppressing disruptions. Many investors express a desire to add positive screens to their current negative filters in order to reward the companies that capably manage environmental risks or that leverage the strategic opportunities of a gradually greening economy. To do that, however, investors need deeply insightful reporting that integrates a company's ECSR metrics with its financial and strategic plans. The ESG data currently demanded by global stock exchanges and collected by influential financial advisory firms typically lack that depth. Professional industry groups have recognized the gap, however, and are working actively to improve ECSR reporting in ways that will enhance investor decision-making.

One improvement already captured in the ISO, GRI, and CASS reporting protocols and gaining prominence in financial environmental reporting is to focus on the most material issues for an industry. Given that environmental risk/opportunity exposure is context dependent, demanding full disclosure from all companies on all environmental indicators has the effect of smothering risk/opportunity signals with meaningless noise. Climate change, for example, poses a lower risk for many professional service companies than for transportation companies whose vehicles emit greenhouse gases (GHG). In contrast, there could be both risk and opportunity for energy companies depending on their mix of renewable and non-renewable resources. Investors would accordingly benefit if climate change reporting was mandatory for transportation and energy companies but optional for professional service firms according to other climate-relevant factors such as location in a high-risk geography. A U.S. accounting think tank—the Sustainability Accounting Standards Board (SASB)—has issued a financial Materiality Map with industry-specific reporting guidelines. Companies following SASB advice can greatly amplify the signal-to-noise information content of their ECSR disclosure reports, thus helping investors to make better decisions.

China's Stock Market Guidelines[3]

China's stock market guidelines state that all companies must explain "significant" environmental incidents that could affect their share price such as accidents, penalties, government designations as a serious polluter, investments with significant environmental impacts, and exposure to adverse regulatory changes. Companies in environmentally sensitive industries like power generation, minerals and metals, and construction materials are subject to the additional *MECSR* requirements of disclosing resource consumption, pollutant discharges relative to national and industry standards, Waste Management, investments in environmental protection technologies and facilities, plus company policies and goals, and results of planned improvements. For those in less sensitive industries, ECSR reporting is optional, "according to the company's own needs."[4]

Overall, China's stock market guidelines are minimally demanding for most companies. They call for less data and disclosure than would be needed by investors to drive decision-making. As discussed in chapter "Introduction", researchers and practitioners have argued that Chinese government officials, rather than investors, are typically the target audience for Chinese CSR/ECSR reports since government policies and practices rather than capital market mechanisms determine the ultimate impact of a company's environmental performance. China's government inhibits NGO protests, consumer boycotts, lawsuits, and other external market pressures that might pose risks for environmental offenders listed on more mature capital markets (Marquis et al. 2016). Accordingly, a Chinese company's weak environmental performance is decoupled from risk ratings that might otherwise suppress stock prices. Similarly, since Chinese capital markets have a relatively low proportion of institutional investors compared to more mature western markets, a company's favorable environmental performance would be unlikely to attract purchase interest from long-run, stable investors in a way that might raise its stock price.

Given the decoupling of environmental transparency and capital market risks and rewards, disclosure often serves a symbolic function in China to signal political legitimacy (Marquis and Qian 2014) to regulators and other interested government bodies. Politically symbolic reports need only demonstrate compliance with government priorities without committing to internal transformations or self-governance. Symbolic reporting might ease a company's access to legal permits or government financing, but unless those advantages can be properly leveraged into competitive skills and resources there would not be a systematic effect on profits (Yu et al. 2017). Accordingly, empirical evidence suggests that Chinese companies subject to *MECSR* reduce their harmful environmental impacts on average, but they do so at a cost to profitability (Chen et al. 2018). Furthermore, it has been shown that investors fail to reward Chinese companies for high-quality ECSR/CSR reporting (Ji 2015; Lyon et al. 2013), unless other company actions are indicative of strategic competency (Zou et al. 2019).

Multi-Stakeholder Reporting

Individual stock exchanges vary in their reporting requirements, but the multi-stakeholder protocols issued by ISO and GRI are standard across geographies (see chapter "Introduction" for an overview of the role of global "multi-stakeholder" organizations in China environmental governance). They are intentionally transformational, designed to facilitate performance tracking and continuous improvement. Though launched with different audiences and usage goals in mind, ISO and GRI have converged over time to support meaningful reporting on the most relevant issues for environmental preservation.

GRI

GRI guidelines were developed by NGOs in close consultation with the United Nations to disclose the social, environmental, and governance factors of greatest impact to a broad base of stakeholders. GRI's voluntary guidelines were converted to verifiable standards in mid-2018. Disclosure areas remain the same in the new standards as those of the preceding G4 general guidelines, but the clarity and quality of the reporting have been enhanced (see https://www.globalreporting.org/standards).

GRI was an early advocate of grounding reports in an analysis of materiality. Initially, however, G4 guidelines were interpreted to mean that materiality is where stakeholder interests and the company's business interests converge. In the 2018 standards, GRI clarified their definition of materiality to prioritize stakeholder impacts without necessarily considering the effect on business results. This strengthens GRI as a governing mechanism since it forces companies to be transparent even on issues that might be strategically or financially difficult to change. GRI also stresses that quantitative indicators must be comparable over time and across companies, and balanced in covering both favorable and unfavorable company impacts. GRI data must also be sufficient for third-party quality assurance, though GRI reports are not formally certifiable.

ISO

In contrast to GRI's emphasis on external reporting to satisfy multiple stakeholders, ISO 14001 guides a company to establish an environmental management system for improving internal operations. The original guidelines emphasize efficiency, waste reduction, and cost control without requiring the firm to set external environmental performance targets. This is modified in the most recent revision (2015), which encourages companies to take a proactive stance on global environmental concerns like climate change, pollution, and preservation of biodiversity. Similar to GRI's emphasis on materiality, the new ISO 14001 standards require that companies consider external stakeholder views when identifying priority areas for improvement. ISO also recommends that companies enhance their stakeholder communications, but the external release of ISO 14001 reports remains optional.[5] ISO 14001 standards are certifiable, which makes the system popular among SMEs and others seeking third-party signals of legitimacy.[6]

A second ISO protocol—26000—is more holistic than 14001, released in 2010 to help a company understand and improve its contribution to sustainable development. Similar to GRI, the approach was formulated with input from hundreds of stakeholders internationally. Environmental protection is one aspect of the guidelines, which also include social responsibility factors like good governance and the cultivation of human and social capitals. Unlike ISO14001, ISO 26000 sets voluntary guidelines that cannot be certified.

CASS

Chinese Academy of Social Sciences, a *shi ye danwei*, has issued reporting guidelines for Chinese companies since 2009 (see https://www.globalreporting.org/resourcelibrary/Linking-CASS-CSR-3.0-and-GRI's-G4-Sustainability%20Reporting-Guidelines.pdf). Over the years the guidelines have been revised to align closely with international reporting standards (especially GRI's G4) while retaining Chinese characteristics such as compliance with national government plans and priorities. Like GRI, CASS version 3.0 advises companies to identify issues that are material to the company and its industry, and to include high-quality quantitative indicators. CASS issues industry-level guidelines.

Data Quality

ECSR reports are most informative to readers when they include quantitative data to track company performance. Data that are standardized within industry, comparable across time periods, and certified by an independent third party such as an accounting auditor or professional consultant give readers the greatest confidence and ability to interpret the report's significance. Investors have stated that they would be more willing to use reported ECSR data in their deliberations if they adhered to these formats (Bernow et al. 2019; WBCSD and PWC 2018). To motivate a high level of self-governance, data ideally would track against an objective target like zero-waste-to-landfill or science-based carbon reductions that establish specific accountabilities for mitigating global warming. By making its future performance goals public, a company reveals the depths of its willingness to tackle difficult environmental challenges, and it signals internal resolve to improve. If the goals stretch the company beyond current capabilities, executives are likely to engage a cross-section of business units to search for the path forward. Well-crafted improvement targets are thus an important element of an effective environmental self-governance system.

Current ECSR Reporting in China

Sample

To get a snapshot of China's current ECSR reporting situation, we analyzed recent CSR reports of SSE- and SZSE-listed companies. At the end of October 2018, we retrieved 783 CSR reports from the year 2016 found on Juchao Information Network's CNINFO website (https://www.cninfo.com.cn/new/index).[7] Finance companies were dropped from the sample because there are no financially material environmental indicators for this sector according to SASB. 57% of the companies in the remaining sample of 718 were listed on SSE, including 88% of those on the SSE Corporate Governance list—required to submit a CSR report—and 80% of those on the SSE SRI Index—recognized by the exchange as exemplars of good practice. The remaining 43% of companies were listed on SZSE, including 82% of those on the SZ100 list (required to submit CSR reports) and 87% of those on the SZSE CSR index (good exemplars).[8] Fifty-six percent of the companies are SOEs.

As seen in Table 1, the sample includes companies from each of the 10 industry sectors specified by SASB in its financial Materiality Map as having environmentally material indicators.[9] Consistent with Chinese stock market requirements, the sample contains relatively high proportions of polluting industries like Resource Transformation, Infrastructure, and Extractives & Minerals Processors. 31% of the sampled companies are subject to *MECSR*, 41% are subject to Mandatory CSR reporting without the detailed ECSR mandate *(MCSR)*, and the remaining 28% of companies are Voluntary CSR reporters *(VCSR)*.[10] Twelve percent of sampled companies issued B or H shares that can be purchased by overseas investors, or they have attracted Qualified Foreign Institutional Investors (QFII) to purchase their A-shares. Table 1 also shows that 53% of the sampled companies referenced one of the recognized reporting protocols in its CSR Report. ISO 14001 was the most commonly referenced—by nearly a third of all firms—followed by GRI, CASS, and ISO 26000. Many of the companies referenced more than one protocol when writing their reports.

Table 1 Sample Composition

Sector	Number	Percent
Total Sample	718	100%
Consumer Goods	69	10%
Extractive & Minerals Processing	87	12%
Food & Beverage	43	6%
Healthcare	52	7%
Infrastructure	122	17%
Renewables & Alt Energy	9	1%
Resource Transformation	171	24%
Service	41	6%
Tech & Communications	61	8%
Transportation	63	9%
Ownership		
SOE	403	56%
Non-SOE	315	44%
Reporting Status	**Number**	**Percent**
MECSR[1]	221	31%
MCSR[2]	298	41%
VCSR	199	28%
Overseas Investors[3]	**Number**	**Percent**
Yes	86	12%
No	632	88%
Protocol	**Number**[4]	**Percent**
None	*380*	*47%*
Any	*338*	*53%*
ISO 14001	231	32%
GRI	131	18%
CASS	123	17%
ISO 26000	58	08%

[1]*Environmentally sensitive industries*
[2]*Includes other SSE Corporate Governance companies, SSE companies with B/H shares, plus SZ100 list*
[3]*Companies with B and H shares, or QFII Investors*
[4]*More than one option can be referenced in a single report, so individual numbers do not sum to the totals*

Data Coding

To assess the substantive quality of each company's ECSR content, we assigned a score for each of six environmental indicators included in SASB's Materiality Map: GHG emissions, air quality, energy management, water and wastewater management, waste and hazardous material management, and ecological impacts (e.g., biodiversity, deforestation, habitat loss). Two independent coders scored the data, and disagreements were resolved by discussion. Cronbach's alpha inter-rater reliability coefficient was 0.83, indicating a satisfactory level of agreement (Nunnaly 1978).

Scores ranged from 0 to 4 as follows:

0 if the indicator was not mentioned in the report;
1 if the indicator was mentioned without including numerical data;
2 if the indicator was mentioned with data that are not benchmarked;
3 if the indicator was mentioned with data that are benchmarked;
4 if the indicator was mentioned with benchmarking data as well as a future goal.

The higher the score, the greater potential for ECSR reporting to motivate internal transformation and self-governance, as a score of 3 indicates progress in environmental performance and a score of 4 demonstrates commitment to a future goal. Scores of 1 and 2 do not indicate the potential for or the past achievement of change.

To understand the extent to which Chinese ECSR reporting is focused on the issues of greatest financial materiality, we also created a composite materiality score for each company in the sample by averaging the individual quality scores *on only the financially material indicators for its specific industry*. For example, producers of meat and alcoholic beverages are both included in SASB's F&B sector, but whereas GHG is a financially material issue for meat producers it is not financially material for alcoholic beverage makers according to SASB. Using industry-specific SASB materiality profiles, we calculated materiality for meat producers as the average quality score of GHG, energy, water, and ecological indicators, and for alcoholic beverage makers as the average on the energy and water indicators.

Analysis

Reporting Incidence

We began the analysis by examining the level of reporting incidence on each of the six environmental indicators. The top line of Table 2 shows the percent of all sampled companies that reported data on the indicator. The bottom of the table shows the range of reporting incidence across 10 sectors. The top line shows strong alignment with current development priorities of the Chinese government. Energy, waste, and water management—subject to binding targets as early as the 11th Five Year Plan (Kostka 2013)—were most commonly discussed regardless of business sector. More than 50% of all companies mentioned these factors, reaching as high as 63.9% in the case of energy management. Consistent with China's 11th and succeeding FYPs and the September 2013 air pollution action plan issued following the early 2013 globally publicized Beijing PM 2.5 "airpocalypse," air quality also was discussed by more

Table 2 Percent of sampled companies reporting on each SASB indicator, by sector

	GHG	Air Quality	Energy Mgt.	Water	Waste	Ecological
Total Sample (%)	23.5	53.6	63.9	57.7	56.1	9.1
By Sector (%)						
Consumer	24.6	50.7	68.1	63.8	65.2	10.1
Extractive	26.4	71.3	70.1	72.4	71.3	18.4
F&B	11.6	60.5	69.8	62.8	55.8	7.0
Health	23.1	67.3	75.0	80.8	57.7	11.5
Infrastructure	29.5	50.8	57.4	50.8	48.3	18.0
Renewable	33.3	44.4	44.4	66.7	44.4	0.0
Resource	25.2	58.5	64.3	64.9	63.2	2.3
Service	7.3	24.4	51.2	19.5	34.2	2.4
Tech & Comms.	9.8	21.3	59.0	32.8	49.2	3.3
Transport	33.3	60.3	65.1	49.2	42.9	6.4
Sector Minimum (%)	7.3	21.3	44.4	19.5	34.2	0.0
Sector Maximum (%)	33.3	71.3	75.0	80.8	71.3	18.4

Black indicates a high incidence of reporting (top 3 of 10 sectors). White indicates a low incidence of reporting (bottom 3 of 10 sectors)

than 50% of companies. However, despite China's Paris commitment to tackling global climate change, and the 12th FYP target for reduction of carbon dioxide use (per unit of GDP), less than a third of companies in sectors other than Renewable Energy and Transportation mentioned GHG (greenhouse gas) emissions in their 2016 CSR reports, perhaps because the reports were issued too soon after the Paris agreement was struck. Ecological impacts were even less likely to merit attention, being mentioned by fewer than ten percent of all sampled companies.

Table 2 also indicates which business sectors were most and least likely to report on each environmental indicator. Table cells were shaded black for the three sectors with the greatest reporting incidence, white for the three sectors with the lowest incidence, and gray for the four intermediate sectors on each indicator. The shading pattern suggests that Chinese companies are sensitive to their relative environmental impacts. For example, the sector least inclined to report on environmental indicators is the low-impact Service sector and the one most likely to report is the high-impact Extractives & Minerals Processing sector which includes many fossil fuel and mining companies plus iron and steel producers. Other patterns are understandable, reflecting higher sensitivity to an indicator by industries with the greatest negative externalities, e.g., Infrastructure's ecological impact and Healthcare's water impact. Nevertheless, there are notable exceptions such as the low incidence of GHG reporting in the Food & Beverage sector, which includes many agricultural companies. Other seeming anomalies reflect China's industry mix; for example, the modest reporting incidence on water management within the Technology and Communications sector reflects the fact that China has few semi-conductor companies which make water an issue of concern for the sector in other countries. Overall, the patterns in Table 2 show that Chinese companies, in addition to following government priorities for reducing energy, air, water, and waste problems across all business sectors, are also aware of more narrowly defined industry-level concerns about the environment.

Non-Reporting

Despite the general level of industry sensitivity, there are still many Chinese companies who neglect to report on *any* of the six environmental indicators though their CSR reports included discussions of social or economic issues of importance to the government. Table 3 shows the total pool of environmental non-reporters to comprise 15% of the entire sample, with the highest incidence in Service, and Technology & Communications companies consistent with their relatively limited environmental footprints. Nevertheless, according to SASB guidelines, every industry has at least one environmentally material indicator with the exception of finance companies, which we've removed from our analysis. We therefore expect a lower incidence of non-reporting even in the less polluting sectors. It is encouraging that *MECSR* companies were less likely than others to be non-reporters on environmental issues (chi-square = 14.0, $p < .001$), yet 9% of such companies still failed to include any of the six SASB indicators in their reports. State ownership (SOE status) did not appreciably lower the incidence of environmental non-reporting on average (chi-square = 0.22, $p < .64$), but having overseas or foreign investors did (chi-square = 88.3, $p < .001$). Only 7% of such companies were non-reporters, as opposed to 17% of companies with only mainland investors. The third factor that reduced non-reporting was adherence to a standardized reporting protocol, which lowered the average to 8% or less depending on the protocol (chi-square = 31.4, $p < .001$), relative to the 22% non-reporting incidence among companies that did not reference a protocol.

Reporting Quality

To get deeper perspective on the transformational potential of Chinese ECSR reporting, we examined the distribution of quality scores for each indicator. A company that reports at quality level-4 would show the greatest potential, given that it tracks progress against targets for future improvement. Unfortunately, Fig. 1 reveals that level-4 reporting is exceptionally rare. A single company (a foreign-owned electronics

Table 3 Companies not reporting on any environmental indicator

Sector	Number	Percent#
Total Sample	111	15%
Consumer Goods	10	14%
Extractive & Minerals Processing	4	5%
Food & Beverage	7	16%
Healthcare	3	6%
Infrastructure	26	21%
Renewables & Alt Energy	1	11%
Resource Transformation	17	10%
Service	14	34%
Tech & Communications	19	31%
Transportation	10	16%
State Ownership		
SOE	53	13%
Non-SOE	58	18%
Reporting Status	**Number**	**Percent#**
MECSR[1]***	19	9%
MCSR[2]	50	17%
VCSR	42	21%
Overseas Investors[3]	**Number**	**Percent#**
Yes***	6	7%
No	105	17%
Protocol	**Number[4]**	**Percent#**
*None****	85	22%
Any	26	8%
ISO 14001	18	8%
GRI	9	7%
CASS	10	8%
ISO 26000	3	5%

[1]*Environmentally sensitive industries*

[2]*Includes other SSE Corporate Governance companies, SSE companies with B/H shares, plus SZ100 list*

[3]*Companies with B and H shares, or QFII Investors*

[4]*More than one option can be referenced in a single report, so individual numbers do not sum to the totals*

#Base values for the percentages are shown in Table 1. For example, 111 of the 718 sampled companies (15%) failed to report on any environmental indicator.

***Significant at $p < .001$

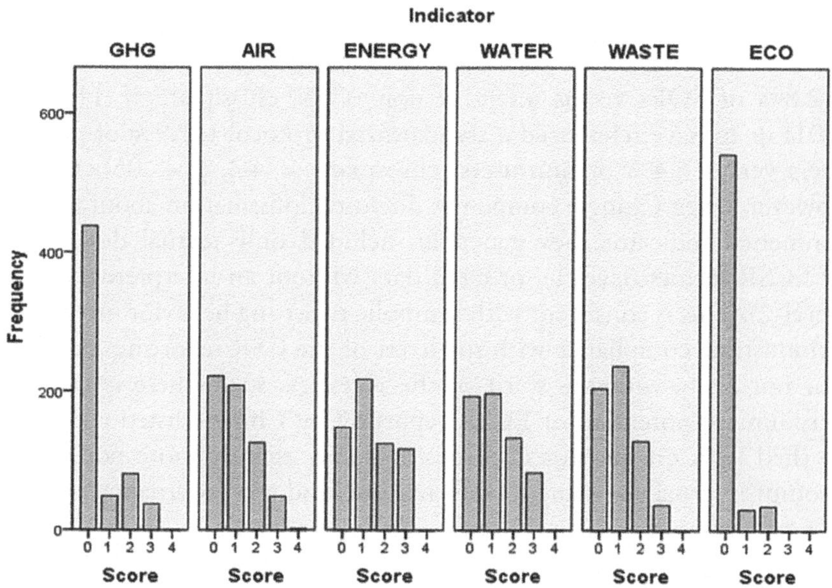

Fig. 1 Quality scores by indicator

producer) earned a score of 4 on the GHG, air quality and water indicators. One other firm (an SOE energy company) earned a score of 4 on only the air quality indicator. No other companies reported at level-4 quality. Overall, it is safe to say that data from Chinese CSR reports shows no signs that companies are setting targets for future performance improvements.

Level-3 reporting (tracking performance data over time) was somewhat more prevalent but still low by absolute standards. 16% of the 718 sampled companies tracked and reported level-3 data for energy, 12% did so for water, 7% for air quality, and 5% each for waste management and GHG emissions. The relatively higher level of reporting quality for energy is consistent with Chinese FYPs, which set binding national targets for energy efficiency as early as 2005 (Kostka 2013). In comparison with Table 2, however, which showed a high percentage of companies mentioning air quality, water and waste management in their CSR reports, Fig. 1 shows that such reporting is generally of limited use for decision makers due to the weak quality of the data.

Companies that reported level-3 or higher quality data on any of the six measured indicators were somewhat more likely to be state-owned (12.8% of SOEs versus 4.6% of non-SOEs, chi-square = 11.9, $p < .001$) or to have referenced a standardized protocol (11.9% of protocol users versus 6.4% of non-users, chi-square = 4.3, $p < .05$). Overall, however, when Chinese companies disclosed information about an environmental indicator they generally included only textual descriptions of ECSR factors (level-1), or cited data without an interpretive context (level-2). This is consistent with symbolic reporting behavior intended to demonstrate compliance with the letter of the CSR reporting mandates, but not really with the spirit of the rules. As such, there is currently very limited potential for ECSR reporting by Chinese listed companies to shed light on investor decisions and only slightly more potential to prompt internal company transformations and self-governance.

Materiality

Finally, we explored the tendency for Chinese ECSR reporting to focus on the most financially material factors. Individual company materiality scores were averaged by sector and other grouping variables, yielding the results shown in Table 4. If a group score is lower than 1.0, it indicates that companies within the group on average neglected to report on factors that SASB considers material to their industry (hence, the average is based on many zero entries) though they might have reported on other non-material factors for their industries. A score that is appreciably higher than 1.0 suggests that the companies within the group tended to focus their reporting on financially material factors, and/or they included data about those factors in their reports. The table shows higher group materiality scores on average for SOEs ($z = 18.3$, $p < .001$) and for those following one of the standardized protocols ($z = 24.0$, $p < .001$). The table also indicates that materiality was directionally higher for *MECSR* companies than for *MCSR* or *VCSR* companies, and for companies having overseas or foreign investors, but neither of those effects was statistically significant.

Table 4 Average materiality scores

	Materiality	
Total Sample	1.09	
Sector		
Consumer	1.49	
Extractive & Minerals Processing	1.08	
F&B	1.01	
Health	0.83	
Infrastructure	0.96	
Renewables	0.75	
Resource	1.22	
Service	1.50	
Tech & Communications	0.91	
Transport	0.85	
State Ownership		
SOE***	1.23	
Non-SOE	0.91	
Reporting Status		
MECSR[1]	1.16	
MCSR[2]	1.10	
VCSR	0.99	
Overseas Investors[3]		
Yes	1.23	
No	1.07	
Protocol		
*None****	0.9	
Any	1.27[4]	
	Latest	**Other**
ISO 14001	1.65	1.15
GRI	1.76	1.11
CASS	1.70	1.38
ISO 26000	1.56	NA

[1] *Environmentally sensitive industries*
[2] *Includes other SSE Corporate Governance companies, SSE companies with B/H shares, plus SZ100 list*
[3] *Companies with B and H shares, or QFII Investors*
[4] *Whole sample average is not a simple average of the sub-category averages due to differences in sample sizes*
***Significant at $p < .01$

Comparing the materiality analysis to Table 2 evidence of sensitivity by Chinese companies to relevant environmental concerns, it is clear that there is broad scope for improvement. It makes sense that companies following a protocol would show the greatest level of improvement, since materiality is a practice that protocols advocate. As discussed earlier, all four of the protocols are sharpening the focus on materiality over time. The bottom right corner of Table 4 provides a breakdown of materiality scores for companies indicating that they referenced the most up-to-date version of each protocol versus those that said they used an earlier version or that didn't state clearly which version was used. Since ISO 26000 has not yet been updated, only a single score is shown. For the other three protocols, however, materiality appreciably increased when a recent version was employed. This is clear evidence that protocols increasingly converge to help companies determine the most important issues for their reporting and improvement efforts.

We conducted one final analysis of the materiality scores to answer the question of whether it helps Chinese companies to reference more than a single protocol. As mentioned earlier, many companies indicated their use of multiple protocols, with some even referencing all four. If companies are merely "ticking boxes" to say that they have referenced a protocol without actually studying the protocol's content or attempting to apply the lessons they've learned, the number of referenced protocols should not influence the observed materiality score. If companies are actually trying to improve in ways that protocols advocate, on the other hand, we might observe a positive relationship between materiality and the number of protocols referenced. To explore this question, we assigned a score from 0 to 4 to indicate the number of protocols referenced by a company. We then regressed the log of materiality against that summed score. A significantly positive relationship was revealed (Model $F = 51.5$, $p < .001$, $R^2 = 0.11$, $\beta = 0.08$, $t = 7.18$, $p < .001$), indicating that Chinese companies are indeed following the protocols in a way that improves the signal-to-noise ratio of their CSR reports. The more protocols a company referenced, the more they focused on materiality and data quality.

Discussion

The influence of China's top-down state planning on environmental governance is evident in the ECSR reporting of our sample. The majority of companies wrote about their efforts to support the long-standing national Five Year Planning targets on boosting energy and water efficiency and reducing air, water, and waste pollution. The state-led push for transparency from heavily polluting listed companies (those subject to *MECSR* guidelines) is clear, as seen in the significantly lower incidence of environmental non-reporting among such companies. It is notable that state ownership per se did not compel companies to address environment concerns in their CSR reports unless they were explicitly subject to the environmental reporting mandate. The tendency of the top-down planning model to instill targeted, rather than universal, accountabilities is apparent (cf. Zhang et al. 2019).

Non-state actors help promote a more generalized acceptance that everyone is responsible for the environment. Accordingly, the companies in our sample with overseas investors and those that referenced the protocols developed by third-party facilitators were less likely than others to omit environmental factors from their CSR reports. The finding substantiates similar conclusions by others (CSR Asia and Embassy of Sweden in Beijing 2018; Hu et al. 2018) and emphasizes the critical role of non-state actors in elevating environmental concerns to the top of Chinese company agendas.

Merely raising the stature of environmental issues is seldom enough, however, to move a company to a state of self-governance. For that, the companies need pressure, training, and encouragement to collect and leverage high-quality data. In this respect, the state planning and control mechanisms are weak, and the non-state influences are not yet reaching their targets. Though SOEs were more attuned to materiality in their ECSR reporting that was true of privately owned companies, the absolute level of the materiality scores remained low. Results were no better for the heavily polluting *MECSR* companies, suggesting that stock market requirements encouraged some degree of symbolic reporting that signals a company's simple compliance with government rules without engendering any deeper sense of accountability for the specific externalities of one's own conduct. Unfortunately, overseas investors were similarly ineffective, and only the guidance of third-party standardized protocols, and in particular the most recent versions of the protocols, was strong enough to have a noticeable effect on the level of reporting quality and materiality. Even then, our sample showed no evidence that firms were benchmarking against future performance targets (level-4 reporting) that would suggest active engagement in self-governance.

Future Outlook

In contrast to market economies where reporting is designed for the information needs of multi-stakeholders, few in China can use the information in ECSR reports to effectively pressure companies to change. The most important stakeholder for influencing Chinese company behavior is the government, and that seems likely to persist in light of recent reaffirmations of centralized authority. Hence, unless top officials prioritize an environmental issue for improvement, it is unlikely to receive widespread attention in Chinese firms.

Fortunately, there are signs that more demanding FYP priorities are on the way. The 2013 air pollution "action plan" was followed by water and soil action plans, and related requirements, replete with demands for companies to absorb the costs of cleanups if they are found to be responsible for problems (*China Daily* 2019). It would not be surprising to see tougher targets put into place with each successive FYP in order to further assist the top leaders in achieving their water and soil quality goals. Moreover, China's commitment under the Paris climate agreement to reach peak carbon emissions by 2030 (NRDC 2015) indicates that GHG emissions will assume much more attention in the near future. In typical step-by-step fashion, China has targeted its initial campaign for carbon control on the Energy and Transportation sectors (e.g., renewable energy infrastructure projects and subsidies for new energy vehicles) without pushing heavily for other sectors to change. To meet the ambitious 2030 targets, much broader cross-sector participation will be required. Hence, we can reasonably expect a wider call and stricter FYP targets for GHG emissions as the 2030 deadline looms near.

Influencing the Disclosure Path to Self-Governance

Even with stronger state-led interventions in setting environmental improvement targets, the need for non-state support will remain strong. The Chinese government is undeniably open to the transfer of knowledge that can help achieve its goals. State-of-the-art reporting protocols that reflect the most rigorous standards and practical insights from global applications meet the government's needs for technology transfer. If stock market guidelines were to adopt more stringent global reporting standards, and financial industry players were to integrate the resulting improved data in their investment decisions and financial product designs, self-governance of Chinese companies would undoubtedly rise. Thus, there is scope for advisory support and lobbying by non-state actors to raise the quality of stock market guidelines and for services that educate domestic professionals about accessing and integrating ECSR data with their operations.

Global bodies like the United Nations and the World Bank's International Finance Corporation, in cooperation with Chinese domestic businesses and think tanks, have issued multiple white papers and reports advising CSRC on the best way to improve Environmental (as well as Social and Governance) reporting guidelines (International Finance Corporation 2011; The People's Bank of China and United Nations Environmental Program (UNEP) 2015). In part due to these pressures, CSRC has issued new guidelines that broaden the mandate for environmental reporting beyond heavy industry and environmental offenders to include all listed companies by the year 2020 (Pan 2018). Moreover, experts anticipate that stock markets will soon issue new reporting guidelines for A-share listed companies that mandate a change from informal and general CSR reporting styles to a more structured and rigorous format that is informative to investors (SynTao Green Finance and China SIF 2019).

Alongside these efforts to influence the regulators, various non-state actors are working to educate and inform financial rating agencies, institutional investors, fund managers, banks, securities firms, and others in the financial community to integrate the newly mandated data into financial product development and investment initiatives. For example,

CFA Institute and UN Principles for Responsible Investment (PRI) recently issued an advisory document that includes case studies that illustrate promising early examples of integration (CFA and PRI 2019), and UNEP Finance Initiative (UNEPFI) joined other organizations including China's Social Investment Forum (China SIF) to organize a roundtable on Sustainable Finance (see unepfi.org/apacrrt) that included informational sessions along with conversations between Chinese stock market regulators, listed company representatives and industry consultants about trends in data reporting and usage. The forum highlighted financial reporting links to climate change and the more general Sustainable Development Goals (SDG), bringing fresh urgency to the issue of reporting quality (UNEPFI, PRI and Syntao Green Finance 2019). Since China has shown strong interest in contributing to both climate control and SDG fulfillment (Xinhua 2016), this persuasive angle may prove to be effective in prompting change.

Influencing the Transformation Path to Self-Governance

As the information *disclosure* path to self-governance evolves within the Chinese financial community, the *transformation* path also gathers pace, aided by efforts from non-state actors like international B2B buyers, supply chain partners, consultants, and industry associations. Buyers and suppliers are demanding ECSR data to meet the informational demands of their own stakeholders. Advisors and professional groups are helping the reporting companies to link their data collection efforts to internal transformation and business strategy development. Going forward, it is imperative for all of these actors to press Chinese companies to meet the highest third-party protocol standards and requirements. Though many of our sampled companies said they reference GRI, a scan of the GRI website showed almost none to be in compliance with the latest standards that demand level-4 reporting. On a more optimistic note, however, the evidence suggests that Chinese companies are in the process of learning from these standardized protocols; hence, any efforts by non-state actors to deepen that learning and link it to objective business gains would be influential.

Both global and local non-state actors are working in this regard. The World Business Council for Sustainable Development (WBCSD), for example, has issued an SDG Compass that helps companies to map the broadly stated environmental performance targets of the UN Sustainable Development Goals to micro-level business plans and financial gains. At the Chinese domestic level, European and American Chambers of Commerce, as well as other industry professional groups, regularly host conferences and competitions that demonstrate best practices in linking data, strategy, and business performance. All of these efforts seem to be having some effect, at least in pointing companies toward the most important issues for their consideration. A consultative analysis of Chinese CSR reports in 2018 showed a notable increase in the materiality and completeness (reporting both favorable and unfavorable issues) of Chinese CSR reports over time, and growth in the level of reporting that meets the information needs of international stakeholders (GoldenBee China 2018).

Other efforts to promote transformation of Chinese companies are embodied in the work of ISO. ISO14001 is a certifiable standard, and our analysis showed it to be the most widely referenced of the four protocols we recognized. The certification is likely a key attraction for companies, as it fits with the general Chinese cultural tendency toward compliance with hierarchical rules and regulations. Though it is weak in driving self-governance through the information *disclosure* effect (since external release of ISO reports remains optional), it strongly feeds into the *transformation* path to self-governance, given the requirements for continuous improvements in environmental management systems.

It is instructive to note that Chinese leaders generally seek pragmatic ways to move forward while limiting risks; thus, the kind of step-by-step, learn-as-you-go change process advocated by ISO is powerful. Encouraging and facilitating the work of Chinese business partners to seek ISO14001 certification and concomitantly, to publicly report on their progress via GRI or CASS multi-stakeholder protocols, is perhaps the most fruitful path of non-state actor intervention. Over time, as ISO and GRI continue to converge in defining best reporting standards and CASS joins in this process, self-governance in China can only continue to improve. Though the reporting path to self-governance might be slow, it heralds continuous progress when well done. China might yet cross the wide river to reach an ecological civilization on the other side. Given the country's recent history of slow starts followed by explosive acceleration on the path to economic development, there is every reason to believe that crossing the stones across China's ecological river will yield similar mid-term results.

Acknowledgements The support of CEIBS Research Assistant Zhang Qiong with data collection and analysis is gratefully acknowledged.

Notes

1. We note that while the CSRC is commonly viewed as a "government agency," it was created as a *shi ye danwei (事业单位)* (see chapters "The State, Nonstate Actors, and China's Environmental Performance: Setting the Stage" and "Will China Industrial Organizations Succeed in Addressing the "Trust Deficit" Arising from Global Supply Chain Governance?"). It is directly subordinated to the State Council and has statutory delegated supervision power over securities markets.
2. Shanghai and Shenzhen stock exchanges, text translated from Mandarin, https://www.sse.com.cn/lawandrules/sserules/listing/stock/c/c_20150912_3985851.shtml, and https://www.csrc.gov.cn/pub/shenzhen/xxfw/tzzsyd/ssgs/ssxxpl/ssplfz/200902/t20090226_95560.htm, accessed November 20, 2018.
3. As noted, the stock markets are *shi ye danwei (事业单位)*, a form of China non-state actor. See chapter "Introduction".
4. Ibid.
5. ISO, *Main Changes in ISO 14001:2015*, accessed November 29, 2018, https://committee.iso.org/files/live/sites/tc207sc1/files/Main%20changes%20in%20ISO%2014001_2015.pdf.
6. ISO, *ISO 14001 Family, Environmental Management*, accessed November 29, 2018, https://www.iso.org/iso-14001-environmental-management.html.
7. Juchao is China's first securities professional website to fully disclose information about SSE and SZSE listed companies.

8. When financial companies are included in the sample, the sample coverage rises to 95% of the Corporate Governance list, 94% of SSE SRI, 95% of SZ100, and 92% of SZ CSR list.
9. SASB recognizes 10 business sectors with material environmental impacts: Consumer Goods, Extractives and Minerals Processing (e.g., building materials, iron & steel, coal, oil and gas), Food & Beverage (e.g., agricultural firms plus restaurants), Health Care (e.g., pharmaceuticals, medical equipment makers, and service providers), Infrastructure (e.g., utilities and property developers), Renewable Resources and Alternative Energy (e.g., fuel sources, paper & pulp, and clean tech), Resource Transformation (e.g., chemicals and packaging companies), Services, Technology & Communications (e.g., semi-conductors, telecoms, software, and IT service providers), and Transportation (vehicle makers and service providers).
10. The sample is divided into three mutually exclusive groups for our analysis: Those who must file a CSR report but have no specific mandate for environmental reporting (MSCR), those who must file a CSR report that includes an environmental section (MECSR), and those who have no specific reporting requirements (VCSR).

References

Bernow, Sara, Johathan Godsall, Bryce Klempner, and Charlotte Merten, 2019. More Than Values: The Value-based Sustainability Reporting that Investors Want. *McKinsey and Company*, August 7. https://www.mckinsey.com/business-functions/sustainability/our-insights/more-than-values-the-value-based-sustainability-reporting-that-investors-want.

CFA Institute and Principles for Responsible Investment. 2019. ESG Integration in China: Guidance and Case Studies. https://www.cfainstitute.org/-/media/documents/survey/esg-integration-china.ashx.

Chen, Yi-Chun, Mingyi Hung, and YongXiang Wang. 2018. The Effect of Mandatory CSR Disclosure on Firm Profitability and Social Externalities: Evidence from China. *Journal of Accounting and Economics* 65 (1): 169–190.

China Daily. 2019. New Law on Soil Pollution will Pinpoint Responsibility, January 2. https://www.china.org.cn/china/2019-01/02/content_74333237.htm.

CSR Asia and Embassy of Sweden in Beijing. 2018. *A Study on Corporate Social Responsibility*, November 27. https://www.csr-asia.com/report/CSR-development-and-trends-in-China-FINAL-hires.pdf.

Eccles, Robert G., and George Serafeim. 2015. Corporate and Integrated Reporting: A Functional Perspective. *Corporate Stewardship: Achieving Sustainable Effectiveness*, 156–171.

GoldenBee China. 2018. GoldenBee Research on CSR Reporting in China 2018 Released, December 27. https://en.goldenbeechina.com/News/show/id/94.

Hofman, Peter S., Jeremy Moon, and Bin Wu. 2017. Corporate Social Responsibility Under Authoritarian Capitalism: Dynamics and Prospects of State-Led and Society-Driven CSR. *Business & Society* 56 (5): 651–671.

Hu, Yuan Yuan, Yanhui Zhu, John Tucker, and Yuxiao Hu. 2018. Ownership Influence and CSR Disclosure in China. *Accounting Research Journal* 31 (1): 8–21.

International Finance Corporation. 2011. Sustainability and Reporting Guidelines Mapping & Gap Analyses for Shanghai Stock Exchange. https://www.ifc.org/wps/wcm/connect/topics_ext_content/ifc_external_corporate_site/sustainability-at-ifc/publications/publications_report_sse__wci__1319577377015.

Ji, Zhenyu. 2015. Market Effect of Listed Corporate Social Responsibility Report Disclosure—Based on Combined Test of the Double Events. *Open Journal of Business and Management* 3: 163–176.

Kostka, Genia. 2013. China's Evolving Green Planning System: Are Targets the Answer? *Frankfurt School of Finance & Management*. Working Paper No. 201.

Kotsantonis, Sakis, and George Serafeim. 2016. ESG Integration in Investment Management: Myths and Realities. *Journal of Applied Corporate Finance* 28 (2): 10–16.

Lyon, Thomas, Yao Lu, Xinzheng Shi, and Qie Yin. 2013. How Do Investors Respond to Green Company Awards in China? *Ecological Economics* 94: 1–8.

Marquis, Christopher, and Cuili Qian. 2014. Corporate Social Responsibility Reporting in China: Symbol or Substance? *Organization Science* 25 (1): 127–148.

Marquis, Christopher, Michael W. Toffel, and Yanhua Zhou. 2016. Scrutiny, Norms and Selective Disclosure: A Global Study of Greenwashing. *Organization Science* 27 (2): 483–504.

NRDC. 2015. Paris Climate Agreement Explained: Next Steps for China. https://www.nrdc.org/experts/barbara-finamore/paris-climate-agreement-explained-next-steps-china.

Nunnaly, Jum C. 1978. *Psychometric Theory*. New York, NY: McGraw Hill.

Pan, Frank. 2018. Revisiting ESG in China: Has Company Performance Improved? *Sustainalytics*, May 30. https://www.sustainalytics.com/esg-blog/esg-performance-chinese-companies/.

Shanghai University of Finance and Economics Centre for Accounting and Big Data Research and China Europe International Business School Centre for Wealth Management. 2018. 2018 Research on CSR Reports of A-Share Listed Companies.

SynTao Green Finance and China SIF. 2019. Top Ten Responsible Investment Trends in China in 2019, February 15. https://www.syntaogf.com/Menu_Page_EN.asp?ID=21&Page_ID=292.

The People's Bank of China and United Nations Environmental Program. 2015. Establishing China's Green Financial System: Detailed Recommendations 14: Make Environmental Information Disclosure Mandatory. https://drive.google.com/file/d/0B1GFkVHt5UUeN3M5a29paFNtOWM/view.

UNEPFI, PRI, Syntao Green Finance, and China UK Pact. 2019. ESG Data in China: Recommendations for Primary ESG Indicators. https://www.unpri.org/fiduciary-duty/esg-data-in-china-recommendations-for-primary-esg-indicators/4345.article.

WBCSD and PWC. 2018. Enhancing the Credibility of Non-Financial Information: The Investor Perspective. https://docs.wbcsd.org/2018/10/WBCSD_Enhancing_Credibility_Report.pdf.

WTO Tribune and Golden Bee Management Consulting Ltd. 2017. Golden Bee Research on Corporate Social Responsibility Reporting in China 2017. https://en.goldenbeechina.com/ueditor/php/upload/file/20171222/1513910252134409.pdf.

Xinhua. 2016. China Releases National Plan for Implementation of UN Sustainable Development Agenda, September 20. https://english.gov.cn/premier/news/2016/09/20/content_281475446661058.htm.

Yu, Hui-Cheng, Lopin Kuo, and Mao-Feng Kao. 2017. The Relationship Between CSR Disclosure and Competitive Advantages. *Sustainability Accounting, Management and Policy Journal* 8 (5): 547–570.

Zadek, Simon. 2018. The Path to Corporate Responsibility. *Harvard Business Review*, December 2004. https://hbr.org/2004/12/the-path-to-corporate-responsibility.

Zhang, Dongyong, Stephen Morse, and Qiaoyun Ma. 2019. Corporate Social Responsibility and Sustainable Development in China: Current Status and Future Prospects. *Sustainability* 11 (16): 1–23.

Zou, Peng, Qi Wang, Jinhong Xie, and Chenxi Zhou. 2019. Does Doing Good Lead to Doing Better in Emerging Markets? An Event Study of Socially Responsible Investment Index Announcements in China. https://doi.org/10.1007/s11747-019-00651-z.

Will China Industrial Organizations Succeed in Addressing the "Trust Deficit" Arising from Global Supply Chain Governance?

Juan Zhang

Introduction

In this chapter, we focus on the evolving role of China business or trade organizations in the development of voluntary sustainability standards (VSS).[1] As we will discuss, in the late twentieth century opening up and reform period these organizations, often transforming from within government organizations, developed with legal status as *shehui tuanti* (social or group organization, including trade or industrial groups or associations), though some with a stage as *shiye danwei* (public service organization).[2] As this chapter will discuss, at this time the most significant development likely is the evolution of institutional infrastructure which may affect and have positive impact on global, as well as China domestic, environmental quality and sustainability.

J. Zhang (✉)
China National Textile and Apparel council, Beijing, China
e-mail: hi_zhangjuan@126.com

D. Guttman et al. (eds.), *Non-state Actors in China and Global Environmental Governance*, Governing China in the 21st Century,
https://doi.org/10.1007/978-981-33-6594-0_5

The developing role of China trade associations follows on the late twentieth century work of global nonstate actors in developing VSS to address the negative impact caused or contributed by global supply chains. As China's role in late twentieth century globalization developed, global efforts by global nonstate actors to control multinational supply chains inevitably and increasingly came to engage China enterprises and their associations.

In 2013, for example, Canopy, a Canadian non-profit environmental protection organization, launched an initiative to reduce the growing negative impact of man-made fibers, including viscose, on ancient and endangered forests around the world (Canopy website: https://canopyplanet.org/campaigns/canopystyle/). Canopy works with forest industry customers, such as book publishers and fashion companies. China is now a big fiber manufacturing and consumption country. Some China fiber manufacturing companies that purchase forest pulp have been blacklisted by Canopy and thus face the challenge of being removed from the supply chain by their international customers. In this context, China Chemical Fibers Association and China Cotton Textile Association responded by collaborating with ten large Chinese viscose producers to develop the Collaboration for Sustainable Development of Viscose (CV Alliance). In March 2018, the CV Alliance launched a three-year Roadmap which claims to explore a way to promote a sustainable viscose supply chain (CV website: https://www.cvroadmap.com/reporten/201812/45.html).

Similarly, a China *shehui tuanti* was prompted into action when, in January 2016, Amnesty International's report "*This Is What We Die For*" disclosed human rights violations in illegal use of child labor in cobalt mining in the Democratic Republic of the Congo (DRC). (Cobalt is used in smartphones and car batteries and other high technology products; the DRC is the largest source of supply). The organization accused the supply chain upstream and downstream companies of nonconformance with *UN Guiding Principles on Business and Human Rights* and *OECD Guidelines for Due Diligence on Mineral Supply Chains from Conflict Areas and High-risk Areas*. Many big international brands (including Apple and Samsung) and a Chinese treatment company invested in the DRC were listed in the report (Amnesty USA website: https://www.amnestyusa.org/

files/this_what_we_die_for_-_report.pdf). Under pressure, international brands requested their Chinese manufacturing suppliers (more than 90% of the global cobalt processing capacity is concentrated in China) to carry out due diligence to mitigate the child labor risk. In November 2017, the China Chamber of Commerce of Metals, Minerals and Chemicals Importers and Exporters (CCCMC) launched a Responsible Cobalt Initiative (RCI) and developed *Cobalt Refinery Due Diligence Guidance* to help upstream and downstream cobalt supply chain companies address child labor risks (CCCMC website: https://en.cccmc.org.cn/news/cccmcinformation/77704.htm).

Globalization affects China deeply, and the rules of global governance also affect China. The voluntary sustainability standards (VSS) developed out of China serve as one of the governance rules in an attempt to mitigate the negative social and environmental impact of the global supply chain in producing and processing countries. China as a world factory has been greatly affected. However, there are great challenges in achieving this ambition. The deep reason is that VSS has not eliminated the "trust deficit" of global supply chain relations arising from the fact that the upstream actors (producers and manufacturers) are not substantially engaged in supply chain governance. Under such circumstances, China industrial organizations began to develop VSS in a different way with the expectation to reduce the "trust deficit" and hence reshape the supply chain relationship.

In this chapter, we will examine the evolving role of China industrial organizations in developing VSS that may apply domestically and, as China goes abroad, globally. We will consider whether and how China industrial organizations are shaping a new foundation and building new mechanisms for VSS which may make contributions to governance across the global supply chain.

Standards and Voluntary Sustainability Standards (VSS)

At the onset, we should note that the term "standard" (*biao zhun*) is global, but specific meanings require local understanding. In the *WTO*

Agreement on Technical Barriers to Trade (WTO Website: https://www.wto.org/english/docs_e/legal_e/17-tbt.pdf), "standard" is defined as a document approved by a recognized body that provides, for common and repeated use, rules, guidelines, or characteristics for products or related processes and production methods, with which compliance is mandatory. It may also include or deal exclusively with terminology, symbols, packaging, marking, or labeling requirements as they apply to a product, process or production method. WTO defines "standard" from the perspective of international trade of products.

In comparison with the WTO definition, in China, "standard" is defined in a way that provides broader scope for standard setting organizations. The China national standard GB/T 20,000.1-2014 Guide for Standardization—Part 1: Standardization and Related Activities—General Vocabulary defines "standard" as a "document, established by consensus through defined procedures, that provides, for common and repeated use, rules, guidelines or characteristics for activities and their results." It is worth emphasizing that, compared with the old version of the national standard GB/T 20,000.1-2002, GB/T 20,000.1-2014, the current definition of standard highlights standard's functions and roles, and removes the requirement of being "approved" by a recognized body [China government] and thereby weakens its nature as a "normative document." In practice, this lowers the entry barrier of standard development, expands the scope of standard-setters, and enables standard to return to its essence as a tool.

The next question is what the voluntary sustainability standard (VSS) concept means in the global and China settings. ISEAL Alliance is a global membership organization for credible sustainability standards. ISEAL Alliance website (https://www.isealalliance.org) defines VSS as "a standard that addresses the social, environmental or economic practices of a defined entity, or a combination of these" (the ISEAL Alliance 2014).[3] Similarly, in this chapter, as we talk about VSS in the China context, VSS came to mean social, environmental, and economic normative requirements applied to enterprises in order to prevent or mitigate the negative impacts of their activities. The VSS we will talk about in this chapter is close to the definition given by ISEAL Alliance. We also note that the China Association for Standardization (another *shehui tuanti*) is planning to develop Guidance for setting voluntary sustainability standards.

VSS in China and the "Trust Deficit" Dilemma

China has acted as world factory and production base for the world market since China began the Reform and Opening policy in 1978. VSS, which aims to prevent and mitigate negative social and environmental impact caused or contributed by the global supply chain in producing and processing countries, links China to the world economy in a new way. China has been affected by VSS for decades. However, with its implementation on the ground in China, VSS has been criticized as green-washing to cover the bloody exploitation of transnational giants. Correspondingly, some producers and suppliers in China are reported to just take temporary actions, even cheating measures, in order to pass audits or get a certificate.

Underlying the criticism, VSS has not eliminated and has even increased, in some cases, the "trust deficit" in global supply chain relations. The concept of "trust deficit" in global supply chain relations usually indicates that it is hard for a group of actors (such as upstream actors including producers and manufacturers as a group, or downstream actors including brands and retailers as a group) in global supply chains to build up trust in one another. The typical "trust deficit" phenomenon or dilemma is reflected in China in the "cat-mouse game" which prevails along with the implementation of VSS developed out of China. Why does the "trust deficit" prevail through the decades? The root cause is that the upstream actors (producers and manufacturers) are not substantially engaged in VSS designed as supply chain governance, in both the VSS development and implementation phases.

The intent of VSS is to drive and move producing and processing countries (China included) from cheap production to responsible and sustainable production. VSS, taking the fashion industry as an example, are designed by international brands and organizations (see Tables 1 and

Table 1 Supplier code of conduct developed by international brands in fashion industry

International brands in fashion industry	The time of supplier code of conduct
Levi's	1991
Gap	1992
Nike	1992
Disney	1996
Adidas	1997
Patagonia	1997
H&M	1997
Zara	2001
IKEA	2001
Uniqlo	2004
Walmart	2005
Jack&Wolfskin	2007
Marks&Spencer	2010
Under Armour	2014

Source Summarized by author

2) to cover a host of issues, ranging from environmental protection to labor rights and working conditions (see Table 3).

Though the intent of VSS seems encouraging, the reality in China shows that the "trust deficit" has been accumulated step-by-step from the VSS development phase to the implementation phase. In the phase of VSS development, almost all the VSS setting organizations publicly claim that their standards development process has adopted a multi-stakeholder approach and achieved a global consensus. However, manufacturing companies or other upstream companies are rarely invited to participate in the development of sustainability standards. This can be observed, for example, from the web site lists of engaged stakeholders disclosed by standard setters.

In the phase of implementation, audit is still the main scheme to put the design of VSS into practice. There are no signs in China that the audit-oriented approach makes a contribution to solving the "trust deficit" dilemma. Downstream actors including brands and retailers expect to promote an audit scheme using their buying power without true care about supplier fundamental improvement. The upstream actors usually allocate resources in responding to multiple audits from various

Table 2 Influential VSS in the Textile and Apparel Industry in China

Category	Headquarters
The first series of VSS mainly focus on labor and work condition	
Business Social Compliance Initiative (amfori BSCI)	Belgium
Social Accountability 8000 (SA8000)	USA
Worldwide Responsible Apparel Production's Principles (WRAP)	USA
The ETI Base Code	UK
The second series focus on the compliance status of raw material and product	
Better Cotton Initiative (BCI)	Switzerland
Global Organic Textile Standard (GOTS)	Germany
Content Claim Standard (CCS)	USA
Organic Content Standard (OCS)	USA
Recycled Claim Standard (RCS)	USA
Global Recycled Standard (GRS)	USA
Responsible Wool Standard (RWS)	USA
Responsible Down Standard (RDS)	USA
Forest Stewardship Council Standard (FSC)	Germany
The third series focus on specific material or substances in the process of manufacturing	
BLUESIGN	Switzerland
Zero Discharge of Hazardous Chemicals (ZDHC)	Netherlands
The fourth series focus on testing result and production condition:	
STANDARD 100 by OEKO-TEX®	Switzerland
STeP by OEKO-TEX	Switzerland
ECO PASSPORT by OEKO-TEX®	Switzerland
The fifth series feature innovative framework or platform with aim to tackle common challenges arising from the evolution of sustainability standards	
Higg Index	Netherlands
The Supplier Ethical Data Exchange (Sedex)	UK
The sixth series is designed to specific requirements of import country	
Customs-Trade Partnership Against Terrorism (CTPAT)	USA

Note This table includes 22 sustainability standards and their headquarters which are most influential in China. Please note this is not an exhaustive list
Source Summarized by author

customers and this is the reason why cheating measures can be easily found and reported. This leads to low efficiency in VSS implementation and a "cat-and-mouse game" phenomenon that has prevailed for decades.

Table 3 Key VSS thematic issues

Area	Topics
Social and Labor	Working hours Wages Discrimination Forced labor Free association and collective bargaining Health and safety Resettlement Gender equality Indigenous people Private security
Economy	Corruption Transparency Trade regulation Intellectual property protection Money laundering
Environment	Hazardous and harm substances Chemical management Water Carbon emissions Wastes Climate change Species diversity

Note This is not an exhaustive list
Source Summarized by author

Under such circumstances, China industrial organizations began to develop VSS in a different way with the expectation of reducing the "trust deficit" and hence reshaping the supply chain relationship.

China Industry Organizations Begin to Develop VSS

When VSS developed outside China began to enter China through multinational supply chains, the initial reaction of Chinese industry, no matter the companies or the associations, was doubt and questioning. For example, domestic media discussed whether "the SA8000 (one well-known international sustainability standard developed by Social Accountability International.) was a new trade barrier" (Tan 2004; Ren 2006, 32). Especially when China joined the World Trade Organization in 2001, many people suspected VSS was a new means invented by developed countries to limit China's economic development and international trade.

Box 1 The Definition of Industrial Organization in China
There is no national legislation on the definition of industry, but some local regulations have definitions of industry organizations. For example, Article 3 of the Shenzhen Special Economic Zone Industry Association Regulations (2014) defines industry associations as "non-profit social organization(shehui taunti) assembled by enterprises in the same industry or across industries, other economic organizations as well as legal persons, managed in accordance with the constitution, and established in accordance with the law."

However, this initial suspicion and pessimism did not stop the pace of VSS affecting China, since sustainability requirements are integrated as one of the terms of commercial contracts between international buyers and Chinese manufacturers. Thus, it is clear that if companies failed to obtain a certificate that the international customer requests or failed to pass an audit they will face the possibility of losing international customers and be unable to enter the international market. Initially, although the China government expressed concerns about the VSS, it did not introduce measures to restrict or encourage them. Thus, these globally initiated VSS could have practical effect independent from China government regulation. Therefore, VSS, coming along with international buyers' business orders, began to affect the internal management system and business practices of manufacturing companies in China.

In the early stage of implementation, many requirements of the international VSS raised questions when put in the context of China's then institutional background and local reality. At that time, many China companies were unable to find a way to meet requirements such as collective bargaining, freedom of association, trade unions, and working hours. When individual companies realized it was a challenge at an industrial level instead of an individual company problem, companies sought technical support from the industry associations.

Thus, China industrial organizations began to carry out dialogue with international VSS setters and international brands and began to develop sustainability standards that were thought to be more consistent with China's institutional background and industry characteristics (see Table 4).

VSS Set by China Industrial Organizations: Replication and/or Innovation?

As of early 2020, more than 20 China industrial organizations have developed VSS. These VSS concern China enterprise domestic issues and activities but also global activities. From experience and lessons learned from VSS developed out of China, central to the answer to the question of their performance is whether the "trust deficit" in supply chain relationships can be reduced through VSS developed by China industrial organizations. In this section, we focus on standard development and implementation—all of which, in turn, shape trust in supply chain relationships.

In the phase of standard development, stakeholder involvement in the standard-setting process, membership design, and technical system adequacy are critical. Wide stakeholder involvement in the standard setting process ensures credibility. Membership design is critical to the arrangement of the role and responsibility of different actors in the supply chain, and the maturity of the technical system affects the effectiveness of standard implementation. These three factors play decisive roles when industry associations act as standard setters.

Following standard development, further factors are key to implementation. These include whether and how the standard is embedded-for purposes of support, incentive, and/or disincentive-in the relevant institutional environment, including in particular China government institutions but also relevant global and domestic industrial and civil society systems.

Table 4 VSS Setting by Five China industrial organizations[a]

Standard-Setting Organization	Name of VSS	About the VSS
China Chamber of Commerce of Metals Minerals & Chemicals Importers & Exporters (CCCMC)	*Chinese Guideline for Social Responsibility in Outbound Mining Investments* (2014) *Cobalt Refinery Due Diligence Guidance* (2017)	The *Guideline* aims to regulate China's outbound mining investments and operations, guide enterprises to develop social responsibility and sustainable development strategies, and establish CSR and sustainable development management systems. It includes scope of application, guiding principles, and covers eight social responsibility issues: organizational governance, fair operation practices, value chain management, human rights, labor practices, occupational health and safety, environment, and community engagement

(continued)

Table 4 (continued)

Standard-Setting Organization	Name of VSS	About the VSS
China Association of Communications Enterprises (CACE)	*CSR Management System of China Information and Communications Industry* (2016)	The *Management System* is a guiding standard by which enterprises in the information communications industry fulfill social responsibility and practice sustainable development. The standard encourages enterprises to specify the scope of responsibilities in accordance with their respective circumstances, integrate social responsibility into corporate strategy and business operations with the support of leadership and resources, and improve enterprises' capacity to mitigate risk and gain opportunity while strengthening management systems, thus enhancing business performance and social responsibility performance. The *Management System* consists of two parts. The first part is a methodology (drawing on the advanced architecture of the ISO management system) and the second part covers eight issues: governance, innovation, human rights, human resources, customer and consumer services, environmental protection, industry cooperation and fair operations, and community relations

Standard-Setting Organization	Name of VSS	About the VSS
Chinese Electronics Standardization Association (CESA)	*Guidance on Social Responsibility of Information and Communication Technology Industry* (SJ/T 16,000-2016)	The *Guidance* is a basic standard to regulate what are the social responsibilities in the information and communication technology industry, and represents a common understanding and action framework of the information and communication technology industry regarding social responsibility. The full text of the *Guidance* describes the basic social responsibility objectives, principles, and practices of the information and communication technology industry, as well as core topics and issues. The industry's social responsibilities include ten core topics (responsible governance, technology innovation and application, employee rights and interests, health and safety, resource conservation and environmental protection, good faith operations, supply chain cooperation, consumer relations, virtual community management, and community engagement and development) and 31 relevant issues
China International Contractors Association (CHINCA)	*Guide on Social Responsibility for Chinese International Contractors* (2010)	The *Guide* includes the scope of application, definition and basic principles, management, and core issues of social responsibility. The core issues include project quality and safety, employee rights and career development, customer (proprietor) rights, supply chain management, fair competition, environmental protection, community involvement, and development. The *Guide* drew on the core ideas of the international community on social responsibility, summarized and absorbed the social responsibility best practices of the industry in recent years, and defined the common understanding of social responsibility for the international contracting engineering industry

(continued)

Table 4 (continued)

Standard-Setting Organization	Name of VSS	About the VSS
China National Textile and Apparel Council (CNTAC)	*Corporate Social Compliance 9000 for Chinese Textile and Apparel Industry* (2005)	*CSC9000T* takes it as its strategic task to guide textile and apparel enterprises in transformation, upgrading, and sustainable development, encourages enterprises to identify social responsibility risks and opportunities in their own products, services, and business relationships, and integrate the risks and opportunities, respectively, into their management systems and business processes by means of due diligence. *CSC9000T* consists of two parts, code of conduct and management system

[a]For more details of China industrial organizations' sustainability standards, visit the following websites. Last modified April 22, 2020: https://www.cccmc.org.cn/tzgg/50622.htm, https://www.csr-cace.org.cn/index.php/standard/index, https://www.cesa.cn/CSR2016/index.html, https://www.chinca.org/cms/html/shzr2/col343/2013-07/01/20130701140751015993087_1.html, and www.csc9000.org.cn

Standard Development

First, standard development and Stakeholder involvement. Stakeholder involvement in the standard-setting process refers to how much expectations and interests can be reflected during the standard setting process (Liang and Zhang 2017). Based on interviews conducted by the author, China industrial organizations have adopted a multi-stakeholder engagement approach, similar to the approach taken by international sustainability standards. China sustainability standards setters emphasize the participation of manufacturing companies, which provides manufacturers the opportunity to engage in the standard setting process, and also provides a means for China companies to express their concerns toward supply chain governance. Almost all international VSS setting organizations publicly claim that their standards development process has adopted a multi-stakeholder approach and achieved a global consensus. However, manufacturing companies or upstream companies are rarely invited to participate in the development of sustainability standards. This can be observed, for example, from the web site lists of engaged stakeholders disclosed by standard setters.

The involvement of upstream actors such as producers and processors in the standard development phase is very common, but there are uncertainties about the recognition and acceptance by downstream actors, such as brands and retailers, of the VSS developed by China industrial organizations. Thus, the involvement of downstream actors is usually considered as a "tipping point" for VSS due to the fact that the international brands and retailers are usually buyers who have the strong power and influence to enable their suppliers to accept the rules developed. It is not an easy task for China industrial organizations to gain support from international brands and retailers. Due to China industry's current position in the global supply chain, although China ranks Number two in the world in terms of GDP, China companies still act as middle or upstream actors, in other words, suppliers in the global supply chain.

However, China may be at a turning point. As illustrated by the rapid evolution of VSS worldwide, a growing number of people realize that root problems usually exist in the end of global supply chain, i.e., in the forests, plantation, mines, smallholder, and family workshops. As such, traceability and transparency are first required to address the roots of supply chain sustainability. For long and complex global supply chains, without the support of manufacturers, it will be impossible to track the components of a final product to the source(s). In this context, international brands (as the ends of global supply chains) are seeking the support of the manufacturers and hence take the VSS developed by China industrial organizations as a potential cooperation platform.

Moreover, there is another trend which is reshaping the position of China in the global supply chain. According to *2018 Statistical Bulletin on China's Outward Foreign Direct Investment*, China's outward foreign direct investment in 2018 was US$143.04 billion, making it the second largest foreign investor in the world. By the end of 2018, more than 27,000 domestic investors in China had set up 43,000 foreign direct investment enterprises in 188 countries around the world, and more than 80% of the world's countries (regions) had Chinese investment. In this setting, company headquarters located in China have to consider how to use VSS as a supply chain management tool to prevent or mitigate the social and environmental risks in the host country. They usually prefer to use the VSS developed by China industrial organizations since they engage in the standard setting process. And with the growing scale of China overseas investment, this will increase the influence of VSS developed by China industrial organizations in terms of geographical range. We note that in some corporate standard setting contexts, consumers and workers are often considered stakeholders for purposes of participation in standard development. This does not yet seem to be the case in global and China VSS standard development (Fig. 1).

Box 2 An example of the China VSS development process: CCCMC Standard Development and Consultation Process[4]

CCCMC organized many stakeholder communication meetings in China and other countries. It organized two formal stakeholder forums to discuss the standard's framework, one in September 2013 and the other in November in the same year, finalizing the standard's outline (framework). Then the initial draft was developed, followed by three

Fig. 1 Gradual Expansion of Stakeholder Engagement. The scope of stakeholder engagement of the five industry organizations mentioned above is gradually expanded (*Source* Summary based on author interviews)

revisions. Opinions were solicited by means of emails, company visits (Aluminum Corporation of China Limited, China Nonferrous Metal Mining (Group) Co., Ltd., and China Minerals Corporation), and workshops. In March 2014, the initial draft was published online for comments and opinions. The draft was posted on CCCMC's official website and the websites of international partners to solicit comments. More than 2,000 opinions were collected. These opinions were not made available to the general public, but disclosed internally in the drafting process work team. They were organized, categorized, discussed at meetings, absorbed, communicated, and given feedback. The feedback was given via emails and meetings; almost every organization, institution, and individual that contributed opinions were given feedback.

So far, however, China industrial organizations have not adopted membership roles in international sustainability standard setters. For example, ISEAL Alliance is a well-known UK-based "multi-stakeholder" organization with the aim of promoting credibility in the VSS process. The majority of global VSS standard setters are members and thus adopt the ISEAL Alliance standard setting code through the multi-stakeholder engagement approach. However, while some China industrial organizations have maintained cooperative relations with ISEAL Alliance, no

China industrial associations have joined ISEAL Alliance. CCCMC is the only China industrial association to reference the ISEAL Alliance Ten Credibility Principles in its standard development.[5] This may indicate that the China industrial organizations want to maintain a certain degree of independence when setting standards.

Second, membership design. Membership design is a key factor in the arrangement of the role and responsibility of different actors in supply chains. Standard setters usually define global supply chain relationships through the design of membership, since the division of responsibilities is potentially defined through membership. In general, membership may be divided into categories which indicate the understandings among members on global supply chain governance.

Box 3 Membership Strategy of Some International VSS

Amfori BSCI provides that "[A]ny business from a small or medium-sized enterprise (SME), to a large company or an association can join Amfori." However, companies whose main activity is manufacturing in defined risk countries are not eligible for any kind of membership. China is evaluated as a High-risk country so China companies cannot apply for membership but they can apply for audit as the supplier of the Amfori members with their permission. (Amfori, headquartered in Brussels, was created in 1977 by European companies to aid in foreign trade; in 2003, the Business Social Compliance Initiative—BSCI—was created to provide tools for social compliance. https://www.amfori.org/content/join-amfori).

Sedex membership includes Brand (A member), Brand and Supplier (A/B member), and Supplier (B member). Different actors in the supply chain need to apply for different membership. Brand (A member), Brand and Supplier (A/B member) mainly use their database platform to manage data around ethical performance, while the Supplier (B member) usually share verification data on the responsible business practices with customers. (Sedex is a UK incorporated Company Limited by Guarantee, which says it is "owned by its members" and reports 60,000 members in 180 countries. https://www.sedexglobal.com/join-sedex/).

Companies who seek to get certified usually need to apply for the membership in advance or get permission by at least one member of VSS (Amfori BSCI, noted in Box 3, is a typical case). Generally, membership categories include manufacturers and suppliers, brands and retailers, and associations. Each category has individual duties and benefits. The member category of brands and retailers usually takes the responsibility to ask their suppliers to get certified or verified. The category of manufacturers and suppliers needs to implement the relevant VSS and accept the audit.

So far most of the VSS developed by China industrial organizations are not based on membership. The only two exceptions are the CV Alliance (Collaboration for the Sustainable Development for Viscose) and the Responsible Cobalt Initiative (RCI). CV Alliance has 14 members, including the China Chemical Fiber Industry Association, China Cotton Textile Industry Association and 12 leading viscose producers, involving nearly 3 million tons of production capacity, and accounting for more than 50% of the global viscose production capacity) (CV website: https://www.cvroadmap.com/Newsen/201901/47.html). Currently, all the members are manufacturers, but CV has aspirations to extend its membership to the downstream supply chain. In 2019, CV Alliance launched the "*Responsibility to Nature-From Forest to Fashion*" initiative and so far, 142 manufacturing companies joined the ambitious initiative (CV website: https://www.cvroadmap.com/membercn/201905/53.html).

The Responsible Cobalt Initiative (RCI), operated by CCCMC, also has membership. With many manufacturer members, RCI members once included downstream companies such as Daimler AG, BMW Group, Contemporary Amperex Technology Co Ltd., LG Chem Ltd., Samsung SDI, Renault Group, and Volvo. This was regarded as a huge breakthrough because upstream and downstream companies can work together to tackle the deep problems in the cobalt supply chain. However, CCCMC delisted these brands from RCI membership because they have not achieved consensus.[6] Based on interviewing with CCCMC, the root cause is that it is very challenging for the actors in different supply chain positions to gain consensus on supply chain governance within one framework. Overall, though there are some cases where upstream actors join as members of China VSS groups, it is still challenging for China industrial organizations to develop an innovative membership arrangement for both upstream actors and downstream actors working within one unbiased framework.

Third, standard development attention to the technical system. The maturity of the VSS technical system, which should be taken into account in the standard development process, is a key to the effectiveness of standard implementation. The technical system generally includes the verification mechanism(s), monitoring, capacity building, and other related measures. The standard setter must develop a well-functioning technical system to verify and monitor the sustainability performance of applicant companies to ensure the standard well-functioning.

Although there have now been nearly three decades of China work on VSS, the technical system is often subject to suspicion and criticism. Suppliers focus on temporary measures to meet an audit, with little attention to basic improvements and sustainability performance. Technical system components including assessments, capacity building and audit schemes can be found in the standard systems developed by China industrial organizations. Based on interviewing by the author with those with standard setting responsibility in China industrial organizations, "continuous improvement" was repeatedly emphasized and expected to be incorporated into the technical system. For example, the CV Alliance (cellular fiber industry) calls for respect for the complexity of reality and the need to develop an improvements roadmap in three stages (CV website: https://www.cvroadmap.com/reportcn/201809/43.html).

Standard Implementation

Here we focus on the relationship of VSS to the surrounding environment as key variables in the success, or not, of standard implementation. These include, most importantly, the China government, but also China civil society, and global actors.

First, there is the relation of China enterprises and industries to the China government, which is central to social direction. In the

China context, VSS came to mean social, environmental and economic normative requirements primarily applied to China enterprises.

The position of China industrial organizations in China's institutional structure is very conducive to the implementation of standards developed by China industrial organizations. In their origins and operation China industrial organizations had and maintain a very close relationship with the China government. CNTAC, (The China National Textile and Apparel Council), is a national federation of textile industry associations. CNTAC was formerly the PRC Ministry of Textile Industry. Following 2001 reform, CNTAC became a *shehui tuanti*. CHINCA was once a *shiye danwei* of MOFCOM and was transformed into a *shehui tuanti*. The institutional advantage of close government relations provides opportunity to China industrial organizations to leverage the China government in the implementation of VSS.

The development of China civil society, as discussed in other chapters, is an important factor which may promote the implementation of VSS. For example, as discussed in this chapter, the Institute of Public & Environmental Affairs (IPE) was founded to "leverage the power of a wide range of enterprises to achieve environmental transformation, promote environmental information disclosure and improve environmental governance mechanisms."

International organizations are also very supportive of China industrial organizations who seek to develop VSS. For example, with reference to standards explained in Table 4 above, interviews revealed that GIZ (the Germany registered development organization whose sole shareholder is the German government) financially supported CHINCA (the China International Contractors Association) in its development of the *Guide on Social Responsibility for Chinese International Contractors*. GIZ also supported CNTAC (the China National Textile and Apparel Council) in promoting CSC9000T in Myanmar, Cambodia, Vietnam and Bangladesh by leveraging their partnership network in Asia. The UK Department for International Development (DFID) provided donor support to CCCMC to develop *Guidance for Sustainable Natural Rubber* (CCCMC Website: https://www.cccmc.org.cn/docs/2017-11/20171107204714430892.pdf) based on their global program "Forest Governance Markets and Climate (FGMC)" (UK Government website:

https://www.gov.uk/international-development-funding/forest-governance-markets-and-climate-2015-grants-round). Interviewees working in China industrial organizations think the reason that international organizations positively support China industrial organizations to develop VSS is a hope that China enterprises will apply these standards and do business in a responsible way. Whatever the deep-seated reasons, the result is that their support is conducive to the implementation and dissemination of these standards.

In this context, global support for China VSS implementation affects and is itself affected by uncertainty about the reality of implementation. For example, international certification companies such as SGS are accredited by CCCMC to serve as monitoring organizations to audit or verify the degree of VSS compliance. However, interviews reveal that even where China enterprises formally adopt a standard, downstream actors are hesitating to recognize the audit or assessment result. This attitude in turn may affect the long-term "buy in" of Chinese enterprises to implement the standards.

In sum, it can be observed from review of standard development and implementation phases and factors that China industrial organizations seek to explore a new way to reshape VSS and respond to the core problem of the "trust deficit" arising from the fact that the upstream actors (producers and manufacturers) are not substantially engaged in supply chain governance. The practice and reality show that the upstream actors have been positively engaged in the standard development and implementation phase but due to the uncertainty in attitude and action of downstream actors, it is still a test for China industrial organizations to succeed in proposing an innovative and inclusive supply chain governance framework.

Conclusion

Following decades of influence from global VSS institutions, China industrial organizations are developing VSS to meet China's domestic context and also contribute to global supply chain governance. Ahead lie challenges due to the current substantial position of China in the global supply chain and China's limited experience in setting modern global trade rules. However, there are also positive signs which indicate hope and opportunity for China companies and China industrial organizations.

The core question, this chapter explains, is how well the factors in standard development and implementation can succeed in addressing the problem of the "trust deficit" arising from the fact that the upstream actors (producers and manufacturers) are not substantially engaged in supply chain governance. From the standard development and implementation phases, we note that the progress is apparent at the level of outputs and outcomes. But the long-term impacts of this effort remain uncertain.

Notes

1. The author notes that she researches and works on industry association standard setting as staff at the China National Textile and Apparel Council (CNTAC). In this context, the analysis in this chapter draws on participant observation, interviews, and tracking the activities of industrial organizations over time.
2. Chapter "The State, Nonstate Actors, and China's Environmental Performance: Setting the Stage" provides background on *shehui tuanti* and *shiye danwei*. In this chapter, industrial or trade associations or organizations are terms used for *shehui tuanti*. Chapter "Non-Industry and Nonstate Actors Contribution in the Standard Drafting Process: Examples from the Development of China Room Air Conditioner Standards" examines the China governmental standard setting process.
3. The ISEAL Alliance. "ISEAL Code of Good Practice for Setting Social and Environmental Standards (V 6.0)." Last modified April 22, 2020. https://www.isealalliance.org/sites/default/files/resource/2017-11/ISEAL_Standard_Setting_Code_v6_Dec_2014.pdf
4. The information is summarized based on the documents provided to the author by CCCMC.
5. CCCMC claims that "(t)he process of developing these Guidelines is geared to the ISEAL Standard-Setting Code of Good Practice for establishing Social

and Environmental Standards, (see Annex 2) Relevant Chinese government departments." See also from: https://cccmc.org.cn/docs/2016-05/201605031614081537 38.pdf.
6. Based on the interview with CCCMC.

References

Amnesty International. This Is What We Die for: Human Rights Abuses in the Democratic Republic of the Congo. Last modified May 18, 2020. https://www.amnestyusa.org/files/this_what_we_die_for_-_report.pdf.

Canopy. Introduction to Canopy-style Initiative. Last modified May 18, 2020. https://canopyplanet.org/campaigns/canopystyle/.

China Association of Communications Enterprises (CACE). CSR Management System of China Information and Communications Industry (Chinese version). Last modified May 18, 2020. https://www.csr-cace.org.cn/Public/ueditor/php/upload/20181113/15420933772702.pdf.

China Chamber of Commerce of Metals, Minerals and Chemicals Importers and Exporters (CCCMC). 2014. Chinese Guideline for Social Responsibility in Outbound Mining Investments. Last modified May 18, 2020. https://www.cccmc.org.cn/docs/2017-08/20170804141709355235.pdf.

China Chamber of Commerce of Metals, Minerals and Chemicals Importers and Exporters (CCCMC). Introduction to Responsible Cobalt Initiative. Last modified May 18, 2020. https://en.cccmc.org.cn/news/cccmcinformation/77704.htm.

China Chamber of Commerce of Metals, Minerals and Chemicals Importers and Exporters (CCCMC). Guidance for Sustainable Natural Rubber. Last modified May 18, 2020. https://www.cccmc.org.cn/docs/2017-11/20171107204714430892.pdf.

China International Contractors Association (CHINCA). Guide on Social Responsibility for Chinese International Contractors. Last modified May 18, 2020. https://www.chinca.org/CICA/info/17113011062011.

China National Textile and Apparel Council (CNTAC). 2005. China Social Compliance 9000 for Textile & Apparel Industry (CSC9000T).

Collaboration for Sustainable Development of Viscose (CV Alliance). Introduction to Three-year Roadmap. Last modified May 18, 2020. https://www.cvroadmap.com/reporten/201812/45.html.

Collaboration for Sustainable Development of Viscose (CV Alliance). Introduction to the Responsibility to Nature-From Forest to Fashion Initiative. Last modified May 18, 2020. https://www.cvroadmap.com/membercn/201905/53.html.

Committee on Social Responsibility of the Chinese Electronics Standardization Association (CESA). 2016. Guidance on Social Responsibility of Information and Communication Technology Industry (SJ/T 16000-2016).

Liang, Xiaohui, and Juan Zhang. 2017. A Study on the Practices and Credibility of Chinese Industry Organizations' Standards Setting. *SJTU International Business Law Review* 2: 44.

Organization for Economic Co-operation and Development (OECD). OECD Due Diligence Guidance for Responsible Supply Chains of Minerals from Conflict-Affected and High-Risk Areas. Last modified May 18, 2020. https://www.oecd.org/corporate/mne/mining.htm.

Ren, Ping. 2006. Actively Respond to SA8000 and Blue Trade Barriers. *HLJ Foreign Economic Relations & Trade* 7: 32–33. https://doi.org/10.3969/j.issn.2095-3283.2006.07.011.

Tan, Haobang. 2004. Against the Blue Trade Barrier—SA8000%. *International Economy and Trade Research* 6: 70–73. https://doi.org/10.3969/j.issn.1002-0594.2004.06.018.

The ISEAL Alliance. 2014. ISEAL Code of Good Practice for Setting Social and Environmental Standards (V 6.0). Last modified April 22, 2020. https://www.isealalliance.org/sites/default/files/resource/2017-11/ISEAL_Standard_Setting_Code_v6_Dec_2014.pdf.

UK Government. Forest Governance Markets and Climate Programme (FGMC). Last modified May 18, 2020. https://www.gov.uk/international-development-funding/forest-governance-markets-and-climate-2015-grants-round.

United Nations Human Rights Office of the High Commissioner. UN Guiding Principles on Business and Human Rights. Last modified May 18, 2020. https://www.ohchr.org/Documents/Publications/GuidingPrinciplesBusinessHR_EN.pdf.

World Trade Organization (WTO). Agreement on Technical Barriers to Trade. Last modified May 18, 2020. https://www.wto.org/english/docs_e/legal_e/17-tbt.pdf.

Extending Enforcement: How the Institute of Public and Environmental Affairs Leverages Public Information to Strengthen Environmental Governance

Kate Logan

Introduction

Within the landscape of environmental governance in China, the Institute of Public and Environmental Affairs ("IPE"), a domestic Chinese non-governmental organization ("NGO"), plays a unique, intermediary role. It operates in a space between those officially tasked with enforcing environmental laws and those who are required to achieve compliance. IPE's more obvious function is as an aggregator: The organization collects government-published environmental information and consolidates it into the Blue Map Database, which can be freely accessed online or

K. Logan (✉)
School of the Environment, Yale University, New Haven, CT, USA
e-mail: kate.logan@aya.yale.edu

Present Address:
Independent Researcher, New Haven, CT, USA

D. Guttman et al. (eds.), *Non-state Actors in China and Global Environmental Governance*, Governing China in the 21st Century,
https://doi.org/10.1007/978-981-33-6594-0_6

via a mobile phone application. As a result, IPE's work has traditionally been framed through the lens of transparency (Lee et al. 2009; Li 2016; Zhang et al. 2016; Wang 2018). Greater disclosure expands the clout of the Blue Map Database as a resource, which in turn enables IPE to push for improved compliance through name-and-shame tactics. However, making existing information easier to access and rearranging it to assess performance is not necessarily the only way—or the most important way—that IPE effects change.

Using the foundation of increased transparency, IPE acts as a civil society regulator that engages its advocacy targets, including local-level governments and corporations, to improve their environmental performance. By developing assessments that rank subjects and incentivize them to interact with IPE to raise their scores, IPE carries out a consultative process that guides local governments and corporations into more consistent compliance. IPE therefore leverages the government's authority to provide a service that ultimately strengthens environmental governance by extending monitoring and pushing for improvements. Those actors who choose not to engage with IPE are punished through low rankings and name-and-shame tactics, while those who cooperate with IPE not only receive higher scores, but also reap additional benefits from improving their performance. IPE thus plays a role in furthering decentered regulation.

By outlining IPE's mechanisms of influence, this chapter illustrates how a specific nonstate actor functions in the "shadow of the state" to participate in environmental governance in China. It begins by delineating the history and mission of IPE as an organization, including its structure and the development of the Blue Map Database, which serves as the foundation for IPE's advocacy efforts. It then clarifies the objective and role of IPE's two most important index ranking systems, the Pollution Information Transparency Index ("PITI") and the Corporate Information Transparency Index ("CITI"), to, respectively, improve government environmental information disclosure and corporate environmental compliance. The following section summarizes IPE's mechanisms of influence within the context of nonstate actors aiming to strengthen environmental governance in China. The discussion analyzes

IPE's impact as related to its mission, advocacy model, and mechanisms of influence. The final section concludes.

Background and Institutional Organization of IPE

IPE was founded in June 2006 by Ma Jun, a Chinese journalist-turned-environmentalist who still serves as the organization's director and oftentimes its public face. IPE's mission is "to promote information disclosure, serve green development, and bring back clear waters and blue skies" (IPE 2019). It is officially registered as a *minban feiqiye* ("nonprofit enterprise") under the Chaoyang District Science and Technology Commission of Beijing City, and its office is currently located in the Jianguomenwai Diplomatic Residence Compound in Beijing. IPE established and continues to operate the Blue Map Database,[1] a resource for environmental data that can be accessed either online or via a mobile phone application (the "Blue Map app"). The database serves as the cornerstone of IPE's advocacy to improve environmental governance and promote the greening of supply chains and finance.

Prior to founding IPE, Ma was a journalist who wrote for the *South China Morning Post* during the 1990s. His investigative reporting brought him in close contact with China's burgeoning pollution and environmental issues and eventually spurred him to publish *China's Water Crisis* in 1999. The book has drawn comparisons to Rachel Carson's *Silent Spring* for channeling the attention of the Chinese public toward the nation's growing environmental emergency, much of which had previously been ignored or actively concealed. Ma spent a brief stint as environmental consultant before being named a Yale World Fellow in 2004. His time at Yale exposed him in greater depth to the history of environmental law in the United States, as well as the important role that information transparency had played in holding polluters to account (Moustakerski 2014).

In particular, Ma was struck by the story of the Toxics Release Inventory (TRI), a database mandated by the Emergency Planning and Community Right-to-Know Act of 1986. The TRI requires entities that handle hazardous materials measuring over a certain threshold to publicly declare information about these substances and their quantities. Its adoption was motivated largely by the 1984 Bhopal disaster in India where chemical releases from a pesticide plant killed thousands upon impact, and a similar smaller incident the following year in West Virginia at a plant owned by the same company, Union Carbide. Following the release of the TRI, NGO efforts such as EDF's Scorecard campaign helped to pressure a number of major corporations to voluntarily commit to reducing their discharges (Fortun 2004).

Particularly notable is the TRI's potential for impact without relying on traditional enforcement mechanisms, such as command-and-control measures, market-based approaches, or the judicial system. The TRI does not require lawsuits or active government monitoring in order to exert pressure. Instead, it acts as a "third wave" regulatory tool that uses information as means of improving environmental performance (Cohen and Santhakumar 2007). For instance, by reconfiguring the TRI's data into a user-friendly format and ranking companies' total emissions, the Scorecard website effectively named-and-shamed polluters into reducing emissions of their own volition. In China, where the status quo of a weak judicial system and inconsistent environmental enforcement seemed likely to persist, Ma hypothesized that disclosure might prove more effective at strengthening environmental governance. Moreover, requirements for active disclosure, whereby entities are required to publish data regardless of whether it is specifically requested, could facilitate nonstate efforts to strengthen governance through name-and-shame and other tactics that had proved effective in the context of the TRI. Hence, the idea for IPE was born: Ma would build a public database of environmental information that could be leveraged for advocacy purposes. That is, the database would, quite literally, provide leverage (see Fig. 1) by expanding information disclosure to make it easier for government, the public, media, and NGOs, including IPE, to pressure corporations into improving their environmental performance.

Early on, another decision was made that would be pivotal to IPE's ability to navigate government sensitivities: The database would not include information that had not already been published or otherwise verified by official sources. Instead of collecting their own data and

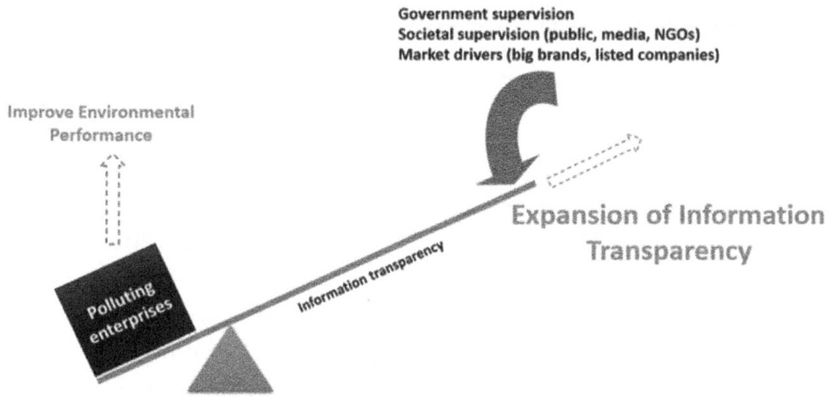

Fig. 1 How IPE provides leverage (IPE 2019)

comparing it to official figures, Ma and his team would collate information from government sources and take care to cite each source.[2] While factors apart from government sensitivities may have played a role in this choice—aggregating government information was already common in other countries, and collecting private data is generally risky—this decision was crucial because it enabled IPE to defer responsibility for the accuracy of information within the database to the government. If companies or other stakeholders were dissatisfied with the quality of the information, they might be motivated to pressure the government itself into improving data quality. Moreover, citing the government as the source of data provided an added layer of legitimacy to IPE's claims when approaching companies. It effectively allowed IPE to turn its claim into a matter of mandatory legal compliance, rather than a mere NGO accusation. By highlighting the gap between law and the current performance of its evaluation targets, IPE would outline a clear pathway to improvement, making its tactics solution-oriented. Furthermore, collecting as much publicly-available government environmental information as possible allowed IPE to not only highlight the range of existing data, but also to identify information gaps that might impede the government's ability to effectively carry out environmental supervision. Limiting the scope of the database to official information ultimately enabled IPE to shield itself from accusations of stepping beyond the state.

Development of the Blue Map Database

The Blue Map Database, which IPE describes as an "environmental data disclosure platform," serves as the foundation for most of IPE's work. While the database was initially developed independently of engagement with the state, corporations, or other environmental governance stakeholders, its later development has become an iterative process, whereby stakeholder interactions have influenced its nature and accelerated its development. The Blue Map Database has seen four major online iterations since IPE's founding, and the mobile application version is currently in its fifth edition at the time of writing.

When the Blue Map was established as IPE's first project in 2006, it was called the China Water Pollution Map and only included water quality data. A second project, the China Air Pollution Map, expanded the scope of the database to also include air quality information. While these two projects focused on environmental quality data, a third stream of data focusing on corporate environmental compliance was soon added. This portion initially consisted of environmental violation and supervision records for enterprises in China that had been cited by official sources for noncompliance with environmental laws. IPE's "Take a Picture, Locate a Polluter" project collected geotagged information about industrial emissions from citizens for the first time, contributing another layer to the database's mapping of environmental compliance. Once reports of pollution were received via *Weibo*, a Chinese platform similar to Twitter, IPE would cross-check the data with the Chinese government's list of "key state-monitored enterprises" ("KSMEs")[3] and upload data to a map that gradually expanded as more citizen reports came in. By 2013, over 4,000 such KSMEs had been located and added to the map. With its expanded scope, starting in 2011, IPE began referring to its collection of information as the "China Pollution Map Database," a name that persisted through the end of the second online iteration of the database.

A major advancement for IPE came with the launch of the Pollution Map mobile app in June 2014. The app's launch was precipitated by a major breakthrough in information disclosure. In the wake of the 2013 "Total Transparency Initiative" launched by 26 Chinese environmental NGOs and led by IPE, which called on the government to require KSMEs to publicly disclose real-time emissions monitoring data, the Chinese government adopted the Measures on Self-Monitoring and Information Disclosure for Key State-Monitored Enterprises (Trial) (MEP 2013). The Measures require any enterprise designated by China's then-Ministry of Environment ("MEP")[4] as a KSME to monitor its air emissions and/or effluent, either manually or via the installation of automatic monitoring equipment. If the latter route is pursued, air emissions values must be published hourly and wastewater quality average values published every two hours. The policy also requires provincial-level environmental protection bureaus to develop online platforms for the public to access this real-time data. This large-scale disclosure of real-time air emissions and wastewater monitoring data for approximately 15,000 major emitters has been lauded as the first of its kind globally (Wang 2018).

This policy motivated IPE to begin collecting all available real-time readings into a single mobile application where any user could check them and post information to social media, specifically *Sino Weibo*, a Chinese platform similar to Twitter. In 2015, the app was upgraded to the "Blue Map,"[5] adding real-time data on air and water quality and changing its name to reflect a more positively-oriented mission. Instead of highlighting pollution, the new name focused on the app's potential to restore clear skies and blue waters. It also reflected a deeper institutional shift at IPE from emphasizing name-and-shame tactics toward effectively collaborating with advocacy targets to help them improve their performance. In other words, leeway was increasingly given to actors where they made good faith efforts to improve.

November 2016 saw a major shift with the launch of the third edition of IPE's online database and its re-branding to become the "Blue Map Database." This re-launch integrated the real-time data in the app with the information originally contained in the online version of the database, and eventually vice versa, since new editions of the app have since been released that include simplified search functions for the environmental violation records.[6] Rather than existing as two separate platforms with different functions, the website and the app have been repositioned as two means of accessing the same, central database operated by IPE.

Within the landscape of environmental governance in China, IPE plays both the role of an "intermediary" as well as that of a direct civil society regulator. The Blue Map Database consolidates information to make it more usable. The information contained within the database is official, government-published data,[7] which would otherwise be scattered across different platforms and difficult to access, sometimes intentionally. IPE hence collects and consolidates data to improve its accessibility (Tarantino 2020). In turn, IPE then uses the collected information to serve as a regulatory watchdog that engages stakeholders—ensuring that the government fulfills its own regulations regarding environmental information disclosure, while also supervising corporations to prompt them into consistently complying with environmental laws and policies.

This is, in essence, what makes the IPE model unique, because it is self-reinforcing. As IPE prompts the government to expand its scope of disclosed environmental information through greater compliance, IPE increases the potential size of its own database, making it more valuable as a means of pushing for corporate environmental compliance and increasing the clout underlying IPE's advocacy. IPE's indices serve as a "nudge" (Thaler and Sunstein 2009) since they prompt and facilitate using the database to strengthen pollution control, without forcing mandatory requirements upon their subjects, as compulsory government policy otherwise would.

Despite drawing inspiration from the TRI, the Blue Map Database also presents several important differences. First, the TRI relies on enterprises' self-reporting of emissions, which are in most cases already compliant with the law. Reorganizing this information to highlight those companies with the highest emissions caters to concerns toward reputational risk, rather than legal risk. The Blue Map Database, in contrast, collects information about violations at noncompliant enterprises to draw attention to them. Where real-time emissions data is collected, enterprises are only flagged when emissions exceed standards. Attention is not actively directed toward those sites that maintain compliance. Second, because TRI data is self-reported, it is oftentimes limited to company's best estimate of its emissions. This makes it possible for enterprises game the system by altering their calculation methodology. IPE's records are conversely based on official monitoring reports and carefully calibrated automatic monitoring systems. Testing methods are standardized so that emissions and monitoring reports exhibit comparability. Although data falsification is still an issue, criminal penalties for tampering with records or monitoring devices were recently strengthened (State Council 2017).

Organization Growth and Current Structure

At the time of writing, IPE employed nearly 40 full-time staff and divides its operations into two main umbrellas: programs and administration.[8] Ma Jun oversees programmatic operations, which include teams dedicated to information disclosure, green supply chain, green finance, energy & climate change, product development (focusing on the front-end operation of the Blue Map app and website), and IT (focusing on maintenance and development of the back-end for the database). Administrative functions are managed by Deputy Director Wang Jingjing and include finance, human resources, office management, fundraising, and communications. IPE also has one "senior researcher" who has technical expertise and deep institutional knowledge of IPE's advocacy methodologies and thus splits her time across a number of different project areas as needed.

IPE's departments employ the Blue Map Database and its components in a variety of ways. IPE publishes reports that fall into six categories, four of which accord with project areas—information disclosure, green supply chain, green finance, and energy & climate change—as well as reports on air pollution ("blue sky road map") and water quality. Nearly all of these reports draw on the government-published information in the Blue Map Database for their analyses, with the vast majority of these taking on the form of indices. The two most consistently-published indices, the Pollution Information Transparency Index ("PITI") and Corporate Information Transparency Index ("CITI"), have been particularly effective at motivating government bureaus and corporations, respectively, to be more transparent about environmental information. Apart from IPE's reports and evaluations, the organization also carries out a variety of other programmatic functions, many of which are reinforced by these assessments. While some of IPE's activities, such as its work on green finance, are not covered in this article, many of these programs adopt similar tactics to the mechanisms outlined. The next section discusses the PITI and CITI in detail to illustrate how IPE leverages government information in its database to highlight and address the gap between policies and practice.

Setting Standards and Guiding Behavior: The PITI and CITI

IPE uses transparency to set standards and guide the behavior of environmental governance stakeholders. Many of IPE's advocacy activities are premised on indices that measure and rank actors' level of environmental information disclosure as well as their responsiveness to related inquiries. In this context, the term "index" refers to a points-based evaluation framework that assesses whether a subject meets specific criteria across a range of categories. The number of points allocated in each category depends upon the extent to which the subject has achieved the criteria in question. Following implementation of the index, evaluated subjects are then ranked based on the total number of points accumulated across all categories. Developing the criteria on which the index is based is a form of standard-setting that motivates actors to achieve higher benchmarks and thus obtain a greater number of points. IPE's indices are generally scaled so that the maximum number of achievable points is 100, thus simplifying the ease of digesting rankings.

A full list of indices developed by IPE at the time of writing is shown in Table 1. The two cornerstones of IPE's evaluation and ranking systems have been the PITI, which measures government environmental information disclosure, and the CITI, which evaluates how transparently corporations address environmental compliance in their supply chains. These two indices were both developed by IPE in conjunction with respective partners at the Natural Resources Defense Council

Table 1 List of indices developed by IPE[a]

Index name	Acronym	Number of annual reports published	Years in which annual report was Published
Pollution Information Transparency Index	PITI	10	2009–2020
Air Quality Transparency Index	AQTI	3	2010, 2012, 2016
Corporate Information Transparency Index	CITI	6	2014–2019[b]
Listed Company Pollution Risk Index	*N/A*[c]	Every two weeks[d]	2015–2016
Corporate Water Risk Assessment Tool	CWRAT	1	2017
Supply Chain Climate Action Index	SCTI	2	2018, 2019
Water Quality Transparency Index	WQTI	1	2019
Water Source Information Transparency and Management Index	*N/A*	1	2020

[a]IPE reports can be accessed here: http://www.ipe.org.cn/reports/NewsReport.aspx

[b]The 2020 edition of the CITI report had not yet been released as of writing. It is typically launched in October

[c]Referred to as the "Pollution List" (*wuran bang* in Chinese)

[d]Unlike other indices, which were generally published once per annum in years published, the Pollution List was an ongoing effort updated every two weeks and re-published in conjunction with the *Securities Times*

("NRDC"), a major American NGO that was one of the first foreign environmental groups establish a program in China in the mid-1990s (NRDC 2020). The PITI was IPE's first index when it was established in 2009. The CITI was first carried out in 2014.

The PITI and the CITI have been consistently published every year on an annual basis since they were first launched. Both rely on a comprehensive process that involves evaluating existing data as well as requesting information and follow-up actions from stakeholders to determine the extent and sincerity of their responses. By creating standards that take government policy as their basis but rely on NGOs for implementation, IPE effectively extends the capacity of government environmental supervision—whether toward local governments through the PITI, or local Chinese industries that comprise portions of the supply chains of major multinational and Chinese corporations through the CITI. Moreover, the interactive, consultative nature of the evaluation process for both the PITI and CITI is a key driver behind IPE's ability to achieve impact, since assessment subjects are encouraged to act to improve their scores.

In line with the technological advancements to IPE's database, IPE is shifting from releasing scores and rankings on an annual basis to maintaining real-time scores. When IPE's website was upgraded in November 2016, a new landing page was created for its green supply chain program that centered brands' CITI rankings. IPE continued to comprehensively review all brand scores on an annual basis, but also began to update marks in real-time if brands made improvements that merited additional points and provided corresponding evidence. This shift enabled IPE to better motivate continuous improvement as compared to only publishing annual scores, since the latter compels targets to be more reactionary. PITI scores have not yet been published online or upgraded in real-time. However, the 10th anniversary PITI report released in January 2020 notes that going forward, IPE intends to reform the PITI to more dynamically assess government disclosure: "Starting in 2020, we plan to expand the evaluation to include more provinces and cities in a more digital way, focusing on reaching wide-ranging populations so that environmental supervision and emission information can better support the green development of China and the world" (IPE and NRDC 2020). Colleagues have corroborated this statement, noting that IPE plans to focus more on the quality of disclosed data, rather than the amount of disclosure.

The evaluation processes for the PITI and the CITI are detailed in the following sections.

The Pollution Information Transparency Index (PITI)

The Pollution Information Transparency Index is an annual assessment of municipal-level environmental information disclosure. The development of the PITI was spurred by two landmark policies regarding information transparency: the Regulations on Government Information Disclosure (the "Regulations") (State Council 2007), and the Measures on Environmental Information Disclosure (Trial) (the "Measures") (SEPA 2007). Both policies were first announced in early 2007 and went into effect on May 1, 2008. The Regulations marked a watershed moment for the authoritarian Chinese government to open itself up to greater public scrutiny. The announcement of the Measures around the same time positioned SEPA, which became MEP in 2008, as an innovator for information disclosure. MEP was the first cabinet-level ministry under the State Council to carry out its own implementing policy specifically focusing on information transparency (Wang 2018). In an attempt to accelerate the implementation of these policies, NRDC and IPE collaboratively developed the PITI, basing its evaluation criteria directly on the new legal requirements. Since the inception of the PITI in 2008, IPE has implemented the assessment on an annual cycle, initially evaluating 113 key cities for environmental protection and eventually growing that figure to 120 cities.[9]

Apart from one major round of revisions in 2014, the framework of the index has remained mostly unchanged. This relative consistency allows for scores to serve as a benchmark of cities' progress or backsliding over time. It has also facilitated cities to adopt long-term planning strategies to improve their disclosure. The initial version of the PITI evaluated each city's disclosure performance on eight different metrics, which are based directly on requirements outlined in existing laws and policies.[10] For each category, four different aspects are considered: systematic disclosure, timeliness, comprehensiveness, and user-friendliness. Each metric is weighted with total possible points adding up to an even 100.[11] As they were developed, the PITI criteria were tested via a sensitivity analysis involving workshops with experts in law, policy, statistics, and governance to ensure that scores and rankings would not be unduly influenced by adjustments to the score weighting (IPE and NRDC 2010). Since the PITI criteria are based directly on the requirements raised in the Regulations and the Measures, IPE is effectively evaluating how well municipalities comply with legal requirements. IPE does not actually assess the quality of the data's content; it rather assesses the quality of data disclosure.

Following the adoption of the Measures on Self-Monitoring and Information Disclosure for Key State-Monitored Enterprises (Trial) in 2014, the PITI was revised to take into account new requirements for real-time disclosure for KSMEs.[12] Two new criteria were added—real-time disclosure of online monitoring information and enterprise annual emissions data disclosure—while the section on EIA disclosure was revised, and the criteria on disclosure of information on enforcement campaigns were eliminated. In addition, a "systematicness-restricted scoring rule" was implemented,[13] preventing cities from achieving higher marks in timeliness, completeness, and user-friendliness than their scores for systematic disclosure.

The evaluation process involves both third-party evaluation of publicly-available information as well as interaction with cities through application requests for environmental information. Once cities are assessed, their scores are distributed to municipal-level environmental protection bureaus ("EPBs"), which are provided an opportunity to raise additional data and feedback for consideration. While some lower-ranking cities complain, it is arguably more common for cities to want to improve. Cities may also receive pressure from provincial-level governments to better their performance. Inquiries from government officials about their PITI scores have led IPE and NRDC to organize workshops with local representatives to discuss strategies for improving their scores. These interactions not only present officials with new ideas for policy innovation, but also play into the competitive dynamic between municipalities, inciting leaders to demonstrate leadership through the PITI rankings. In other cases, back-and-forth with officials might lead IPE to effectively "coach" the government toward better governance. The PITI thus blurs the distinction between third-party evaluation and collaborative governance by opening up channels for interaction between IPE and government officials. IPE operates as an independent evaluator and makes the final judgment on scores, but because it adopts a consultative approach, it is not fully third party. This is not unlike many ESG evaluations that are objective yet designed to bring about improvements through recommendations.

Finalized PITI scores are consolidated and published on an annual basis in a report that is accompanied by a press release conference. The report caters to local officials' penchant for competition by adopting name-and-shame tactics that highlight the performance of leaders and laggards. The release conference has typically been held at the Publicity and Education Center of the Beijing Environmental Protection Bureau and includes environmental officials from the Ministry of Environmental Protection ("MEP") as well as local-level officials, academic experts, and NGO representatives. Hence, the report and rankings are not necessarily the ultimate products in the evaluation process: IPE intends to engage key stakeholders in an ongoing dialogue surrounding the improvement of government environment information transparency in China. To this end, each report clearly highlights areas of progress and key shortcomings, and also includes a series of recommendations as to how government stakeholders can improve the level of environmental information disclosure going forward. These recommendations, which are typically aimed at the central government, often seek to strengthen legislation in order to expand the scope and comprehensiveness of environmental information disclosure. However, since the reports include a range of case studies highlighting innovation and best-case practices among local governments, they functionally serve as a manual to guide provinces and municipalities into improving their own disclosure practices.

The results and impact of the PITI are evaluated both quantitatively and qualitatively. Quantitative considerations include changes in the scores of cities included in the assessment scope and the number of locations disclosing certain types of information. Qualitative analysis looks at changes in environmental governance, such as improvements in the accessibility of cities' platforms for information disclosure, adoption of new policies that stipulate a legal basis for disclosure of new types of information, and greater central government efforts to guarantee the enforcement of existing laws and policies. The PITI reports emphasize the scientific nature of the assessment by focusing foremost on quantitative changes to cities' scores and rankings, as well as fluctuations in the average PITI score. The relatively consistent nature of the evaluation criteria allows for these results to be compared over time. Perhaps more importantly, the evaluation criteria facilitate more granular analysis of which geographies are complying with which ordinances on environmental information disclosure, and to what extent they are conforming. By exposing which local governments are not following central-level policies, the PITI addresses challenges of local protectionism that might otherwise hamper environmental information disclosure efforts. As a third-party evaluation, the PITI's combination of standard setting and behavior-guiding helps to bridge the gap between policy adoption by the central government and enforcement by local governments, thus strengthening the overall state of environmental governance.

The Corporate Information Transparency Index (CITI)

The Corporate Information Transparency Index ("CITI") evaluates how transparently and comprehensively multinational and Chinese companies manage the environmental performance of their supply chains in mainland China. Similarly to the PITI, it was developed in conjunction with NRDC. The CITI was officially launched in 2014. Its origins stem back to a series of investigations into the information technology ("IT") industry in 2010 and the textile industry in 2012, whereby IPE systematically assessed the responses of corporations to suspected instances of environmental noncompliance in their supply chains. Allegations were founded on public information linking environmental supervision records in the Blue Map Database to the supply chains of specific brands.[14] The CITI essentially took the framework that had been developed for these investigative reports and codified it into an index that could be applied to a corporation in any industry based on the same standardized process. The CITI ranks brands and corporations with supply chains. Although most evaluated companies are consumer-facing, some suppliers are included where they have engaged with IPE regarding their own supply chains. Examples of this include Foxconn, a major supplier to IT companies such as Apple, and Esquel, a textile manufacturer that supplies to several major brands. IPE has expanded the CITI in recent years to include some industries that interact less directly with consumers, such as industrial chemicals and environmental and waste management.

The framework of the CITI has been revised every year since its inception, and a report and rankings have been issued on an annual basis. Despite numerous changes to the details of the CITI, the overarching structure has remained consistent over time. The CITI 1.0 criteria launched in 2014 consisted of five main categories meant to serve as a "green supply chain road map" for companies to improve their supply chain environmental management: communication and follow-up; compliance and corrective actions; extend green supply chain practices; data disclosure; and recycling and reuse. Each of the ten total sub-indicators divided across these categories contained five levels, with the criteria weighted such that the total possible score is 100 points, similarly to the PITI evaluation. The annual revision of the criteria is a consultative process: Experts and brands are contacted and invited to provide feedback to IPE in writing and/or via conversations. Comments are summarized and considered; however, decisions on modifications are made collaboratively by IPE and NRDC, with Ma Jun oftentimes making the final call. Apart from ensuring the criteria are sufficiently achievable to maintain brands' respect for the index, the consultative nature of this process also deepens brands' understanding of the criteria and can influence their will to improve their practices accordingly. These five main categories have seen minor changes over the course of the CITI's implementation, with the 2019 CITI report, respectively, replacing categories one, four, and five with responsiveness and transparency, energy conservation and emissions reduction, and promote public green choice. The scope of the index has also grown, from 147 brands across eight industries in 2014 to 438 brands across 19 industries in 2019 (IPE 2019). While scores have perhaps shifted due to the alterations in indicator weighting and criteria, the same idea of a progressive green supply chain road map has been maintained, allowing for brand rankings to pose some degree of comparability over the years.

The evaluation process for the CITI depends on information actively disclosed by the government, as well as information voluntarily published by companies and provided by them upon request. For companies subject to evaluation for the first time, the initial step is to send the company a letter requesting a response to a series of questions about environmental violations at a suspected supplier or suppliers. The letter includes questions about whether the company maintains a "supplier code of conduct" or similar policy requiring all suppliers to maintain environmental compliance. If so, then IPE bringing a violation to that company's attention should theoretically assist that company in implementing this voluntary standard. In practice, however, many companies dodge responsibility by arguing that the supplier does not have a direct contract with the brand—even though it may be a sub-supplier to one of the brand's direct suppliers. If no response is received, a second letter is sent. If the company responds, then IPE and other NGOs in the Green Choice Alliance ("GCA")[15] will engage the company to explain how the company should screen its supply chain for environmental compliance issues and request suppliers with violations to publicly respond, or undergo a supervised third-party audit to "de-list"[16] any records from the Blue Map Database. The requested follow-up actions are all codified in the CITI, with the company's score reflecting the extent to which the brand engages in this follow-up process. This framework is an example of how IPE uses "nudging" tactics to spur improvement without having the express authority to implement compulsory requirements.

Apart from this evaluation process, several other factors motivate companies to work with IPE. First, since before the launch of the CITI, IPE had already launched investigations into the linkages between major multinationals and environmental violations at China-based suppliers. These reports often adopted name-and-shame tactics that drew attention to laggards. Heavy-hitters were singled out, which led to some early wins for IPE—notably Apple, which reversed its relative silence and committed to substantive improvements at the end of 2011 after a targeted campaign beginning in 2010. The wins from these campaigns, which primarily focused on the IT and textiles industries, meant that IPE possessed a small but critical mass of brands that were already utilizing IPE's resources for supply chain management when the CITI was launched in 2014. Although IPE has continued to call out specific brands and industries through targeted investigations after the CITI's launch, these later campaigns have generally been smaller efforts that have seen mixed and/or relatively short-lived results. With a lack of capacity to research all brands in the same way that IPE focused its efforts on Apple and other early wins, the consultative nature of the CITI and the corresponding value-add of IPE's resources have become more important. Two benefits discussed in detail later in this paper include cost and other savings for brands that can more effectively target audits, and preventing environmental violations from disrupting supply chain operations. Finally, for strong performers, IPE goes beyond CITI rankings to highlight these companies in other ways, such as by adding a "CITI Master" category to the most recent iteration of the CITI (IPE 2019).

Despite the centrality of transparency to its name, the CITI arguably aims more to promote corrective actions and proactive management. In fact, over half of its points measure how companies use disclosed information to regulate compliance and corrective actions among both direct and sub-tier suppliers. Obtaining a high score requires for that brand to be willing to actively engage its suppliers and take actions to resolve any violations. More points are allocated to brands whose suppliers verify corrective actions by undergoing third-party audits supervised by NGOs, rather than just disclosing information about corrective actions. The CITI focuses on motivating substantive improvements, instead of penalizing companies for the volume of issues that initially arise. It is thus entirely possible for a brand to receive a high score if that company has many suppliers with environmental violations, so long as that company is willing to comprehensively motivate all of those suppliers to clean up. The idea is that transparently resolving violations theoretically reduces the occurrence of repeat offenders and simplifies management for brands in the long run.

To facilitate improvements, the CITI is set up to nudge brands into engaging with IPE's resources. It is difficult to receive a high score through actions that do not involve direct interaction with IPE and the Blue Map Database. This structure is premised in part on the notion that the Blue Map Database is the most comprehensive publicly-accessible resource of enterprise environmental compliance and emissions data for mainland China. If companies are looking elsewhere to monitor and guarantee their suppliers' environmental compliance—as is often promised in supply chain codes of conduct—then it is likely they are missing information that is included in the Blue Map Database, but overlooked by these other potential resources. In other words, the CITI is set up such that a brand could hypothetically take the requested actions in a number of different ways—and companies are offered the choice of doing so—but the most convenient, comprehensive option in most cases is to fulfill the requirement by using the Blue Map Database.

The CITI report is released annually at the "Green Supply Chain Forum," an event typically co-organized with government-affiliated entities and other NGOs. Within the report, the results and impact of the CITI as an advocacy mechanism focus on progress for two different stakeholder groups: brands, which are ranked on the CITI, and suppliers, which are not ranked but whose actions contribute to brands' rankings. For brands, changes in rankings are measured from year to year and analyzed according to different groupings, such as industry and region. Brand scores are also listed, but not analyzed in detail due to lack of comparability between different versions of the CITI criteria. For suppliers, progress is measured by the total number of enterprises that have publicly responded to government-issued environmental supervision records and/or disclosed environmental data. The extent to which suppliers have responded is divided into those that have provided public explanations toward their environmental violation records and those that have undergone supervised GCA audits[17] to verify rectification efforts. The number of suppliers voluntarily disclosing their annual emissions data via IPE's "pollutant release and transfer register" (PRTR)[18]platform is also highlighted.

These latter indicators serve as a proxy to represent the extent to which environmental issues have been sustainably rectified—that is, assuming that an enterprise makes a public commitment to corrective actions and/or verifies the effectiveness of these corrective actions, then that enterprise can be taken to have "cleaned up." This process of engaging with IPE to disclose information resolves issues with the cost of government violations being too low to incentivize consistent compliance, since many enterprises would prefer to pay relatively cheap fines rather than pay for more expensive equipment or other solutions that would prevent the issue from re-occurring in the future. The CITI reports also aim to influence other standards by making recommendations directly targeted at different relevant stakeholder groups, including government, Chinese enterprises, financial institutions, and third parties, including media and NGOs.

Mechanisms of Influence

If IPE's publicly-stated mission is "to promote information disclosure and serve green development in order to restore clear waters and blue skies," how does IPE set its agenda? And what factors shape the consultative nature of its activities—especially the implementation of the PITI and CITI indices? This section examines the variety of roles that IPE plays within the ecosystem of environmental governance in China. It adopts the framework presented by Guttman et al. (2018) and summarized in Chapter "The State, Nonstate Actors, and China's Environmental Performance: Setting the Stage" of this book to contextualize IPE's activities as they relate to the streams of influence for nonstate actors participating in environmental governance processes in China. Channels for direct and indirect influence are discussed. Many of these mechanisms reflect converging interests in environmental disclosure among state, society, and international actors discussed by Wang (2018) and their corresponding social effects in enabling advocacy. However, this chapter builds upon the previous work of Wang and others by detailing exactly how IPE leverages transparency as a regulatory tool to also consultatively engage actors and motivate improved performance.

In contextualizing IPE's streams of influence, a few important aspects should be considered. First, as an officially registered *minban feiqiye,* IPE is a nonstate actor. Since it is not a government-organized NGO ("GONGO"), the organization is not required to directly respond to requests of the state and has relative freedom in setting its own operational priorities and direction. At the same time, maintaining organizational existence requires navigating sensitivities toward civil society that are unique to the China context. This is in part why the Blue Map Database contains only data gathered from official sources. Moreover, IPE's existence is reinforced by aligning its activities with government priorities and its role in supervising and advancing the local implementation of central-level environmental policies. Of course, this relationship is more complicated than pure alignment, since IPE arguably also influences the direction of government priorities through its own activities and standard-setting.

Another key aspect to note is that IPE's activities are self-reinforcing. On the one hand, as IPE engages with local governments to promote compliance with national-level environmental information disclosure requirements and optimize the manner in which information is provided to the public, IPE achieves the "promote information disclosure" portion of its mission. At the same time, it also creates resources that can be leveraged to "serve green development" through IPE's green supply chain and green finance work, which will contribute to restoring "clear waters and blue skies." IPE hence designs its agenda with scalability in mind: as an index or standard is developed, it is often done in such a way that its scope of influence can be later expanded. For instance, instead of the CITI being industry-specific, is generalized so that the scope of evaluated industries and companies can continue to grow. The implementation of green supply chain practices can also be extended geographically in China through participation by organizations in the Green Choice Alliance networks of NGOs. The same goes for the PITI, as partner organizations use the framework to evaluate additional municipalities in their local jurisdictions. Making evaluations dynamic, such as by updating CITI rankings in real-time, further enables IPE to scale up its influence. These factors all play a role in both agenda-setting and the ultimate effectiveness of IPE's work. They also pose implications for the degree of replicability of IPE's model and the potential shape of future activities outside of China.

Direct Influence

Implement Government-Set Standards and Ensure Compliance

Supporting the implementation of government-set standards is perhaps the most obvious function of IPE's activities. The PITI is explicitly set up to do just that: The criteria are based entirely off of government-issued laws and policies, and IPE aims to evaluate the extent to which each municipality included in the evaluation is implementing these policies. Part of the idea is that due to fragmentation (Lieberthal 1997), the central government does not necessarily possess the capacity and/or the motivation to ensure that these policies are enacted on a local level. Given signals by central authorities toward prioritizing environmental information disclosure and governance, such as through the 2008 Measures on Environmental Information Disclosure, over time the PITI has evolved to become not only tolerated but even welcomed by government officials at both the central and local levels. This attitude is reflected, for instance, by the annual participation of central and local authorities in the PITI release conference, as well as trainings and other activities that IPE carries out in conjunction with officials. Local governments can also benefit from improved environmental enforcement through stronger performance on cadre evaluations (Kostka and Mol 2013).

Of course, it is still not the case that the PITI is welcomed by local governments across the board, as some officials continue to prioritize economic growth by supporting polluting industries over environmental protection efforts. IPE thus serves as an extension of the central government's capacity by monitoring and pushing to close the gap between central-level policies and their oftentimes inconsistent local-level enforcement. It is important to acknowledge the co-benefit in that by implementing the PITI and pushing for more comprehensive and convenient implementation of environmental information disclosure, IPE also reinforces the influence of the CITI and Blue Map Database by increasing their scope and value for companies.

"Shadow of the State" Influencing Agenda-Setting

The PITI and CITI are both examples of how IPE operates in the "shadow of the state" by harnessing the government's authority to influence actors (Guttman et al. 2018). The PITI follows existing disclosure laws to evaluate and promote improved local government compliance. The CITI relies primarily on information that has been disclosed as a result of the policies tracked by the PITI to motivate brands to improve environmental compliance in their supply chains. In doing so, it leverages economic clout to transform multinational brands into quasi-regulators that hold their supplier factories accountable for violations of Chinese environmental law (Wang 2018). A factory approached by its brand customer and asked to publicly respond to a violation faces two choices: comply and improve its environmental performance, or refuse to comply and risk incurring economic losses that may threaten its survival. Multinationals are, in turn, accountable to IPE, who evaluates their performance as quasi-regulators through the CITI. Factories that publish their corrective actions and open them to public scrutiny may be motivated to maintain environmental compliance. Hence, through a web of indirect interactions, the CITI plays the role of a local supervisory agency toward those enterprises with environmental violations already identified and publicized by the government.

The importance of maintaining government tolerance toward IPE's license to operate also plays a role in agenda-setting. In line with strategies discussed by O'Brien (2013), IPE adopts tactics that further the commitments of the state and seek government contact, but also aim to influence its future agenda. One example of this is IPE's consistent collaboration with government actors for the press release conferences of the annual reports for both the PITI and the CITI. The relationship in these situations is mutually beneficial: By seeking direct government participation in the conference but not the report, IPE can gain de facto approval of the contents of the report and ensure that the government is aware of its activities. On the flip side, government presence also serves to reinforce IPE's clout as a nonstate actor, both toward local governments with the PITI, and in the eyes of corporations and their supplier factories for the CITI. While IPE ensures that its own activities are in line with government priorities to enough of an extent to ensure that this type of collaborative relationship can continue—e.g., by actively citing government policy advancements and campaigns in its reports—the government's exposure to IPE's activities and recommendations arguably influences its future policy direction and likelihood of supporting the types of policies for which IPE advocates.

IPE also influences government agenda setting by channeling disclosure to garner public support for policy innovation. This strategy, which exhibits similarities to the discussion of nonstate actor involvement in agenda setting by Kingdon and Stano (1984), is reflected, for instance, in IPE's role in advancing real-time data disclosure for KSMEs. The actions of the 2013 Total Transparency Initiative ("TTI") which called for required disclosure of real-time emissions data from KSMEs likely played a role in the government's decision to promulgate the Measures on Self-Monitoring and Information Disclosure for Key State-Monitored Enterprises (Trial), which went into effect on January 1, 2014. From a longer-term perspective, the development and implementation of the PITI enabled the TTI to make this recommendation based on evidence of increased environmental information disclosure. The 2008 Regulations on Government Information Disclosure and the Measures on Environmental Information Disclosure (Trial), in turn, facilitated the creation of the PITI. This chain of events demonstrates the interrelated, cyclical nature of IPE's activities and government priorities, whereby IPE's efforts to strengthen implementation enable the adoption of stronger laws and policies, which in turn generates a new round of efforts.

Provide Resources Essential to Enterprise Activity

The Blue Map Database consolidates access to over 1.7 million government-issued environmental supervision records, making the search of these records simple and user-friendly. The labor required to locate, acquire, and process this data, which are heterogeneous and scattered across different sources, is intensive and constitutes a major value-add of IPE's (Tarantino 2020). The database also provides a platform for companies to take a variety of voluntary actions to increase supply chain transparency, such as by disclosing their suppliers on a brand supply chain map[19] or publishing annual emissions data on the PRTR platform. An even newer tool launched in 2017, Blue EcoChain, enables companies to receive coordinated real-time push alerts when new violations are added or supplier factories act to respond to their violations. All of these resources are helpful means for brands to enforce their commitments to supply chain environmental compliance, as well as make good on their pledges to disclose and reduce their carbon and environmental footprints. For supplier factories, the Blue Map Database provides a platform to transparently demonstrate through information disclosure that environmental violations have been resolved and consistent compliance has been achieved.

The operational value of the Blue Map Database and Blue EcoChain is important because IPE does not have the capacity to deploy targeted campaigns to name-and-shame the several hundred companies included in the CITI. The assessment therefore functions as a nudge to alert corporations to the added value provided by IPE's resources for supply chain management. There are two main avenues through which these resources support companies' operations. First, IPE's resources help reduce costs and capacity required for active supply chain monitoring. Rather than sending auditors to large numbers of factories, which is expensive and takes time, companies can identify the riskiest suppliers for on-site audits. Furthermore, Blue EcoChain's push alert mechanism reduces the need for staff to constantly monitor suppliers; rather, issues are automatically raised as soon as they arise. Second, they enable companies to proactively mitigate the risk of environmental-related shutdowns and other supply chain disruptions. This function has become more valuable as enforcement ramps up: Beginning in 2016, the central-level environmental authorities in China launched a series of targeted campaigns to enforce environmental laws on a local level. Using the Blue Map Database enables brands to identify potential problem suppliers and resolve issues before factories are shut down by the authorities, thus preventing potential supply chain disruptions. These resources support the operational needs of brands and supplier enterprises alike and reduce their need for internal capacity.

Provide Expertise that Companies Value

The CITI is intentionally designed to be a "green supply chain road map" that lays out a framework for how corporations can improve their supply chain environmental management. Its design does not punish companies for the existence of violations but rather rewards them for comporting with recommended follow-up actions for ensuring compliance, extending supply chain management, and voluntarily disclosing data. Implementing the index and ranking brand performance incentivizes companies to take advantage of this expertise. Many companies initially contacted by IPE regarding violation records in their supply chains are completely unaware of the existence of these violations and/or how to go about resolving them. Because IPE offers both the resource of the Blue Map Database and the institutional expertise on how to engage with it through the CITI and a variety of other guidances and resources on IPE's website, companies' environmental management directly benefits from actions to improve their CITI scores. Both "carrot" and "stick" mechanisms are therefore at work: The risk of negative publicity from a bad score serves as a "stick," while the "carrot" of praise for a high ranking and better overall management provides incentive for brands to engage with IPE. One colleague notes that IPE's ideal relationship with brands is not so much collaboration, but rather a "dynamic tension" whereby both parties agree upon the end objective but maintain different forms of leverage. IPE could reduce the company's CITI score or even launch a targeted investigation if it falls behind, but is also happy to praise the company and recognize its progress if it meets certain performance conditions. For IPE, the co-benefit of the resultant brand participation is the cleaning up of pollution in China through supplier actions to rectify environmental violations.

Indirect Influence

Incentive of Going Abroad

Among the 306 brands evaluated in the 2018 CITI report, 142 are from the Greater China region. The report notes that these brands have traditionally scored lower than brands from other regions. This is due in part to a lack of sensitivity to consumer pressure, but also other constraints, such as lack of internal resources or need to comply with environmental disclosure regulations for companies listed on international exchanges (IPE and NRDC 2018). Increasing desire of companies to expand their operations outside of China poses the potential to change this status quo. For instance, in 2018, IPE launched a campaign highlighting Xiaomi's failure to disclose environmental risks in its supply chain ahead of its IPO on the Hong Kong stock exchange (IPE and Lvse Jiangnan 2018). Although the campaign ultimately saw mixed results, it received attention from international media (Feng 2018; Standaert 2018) and highlighted the increasing risks of private Chinese enterprises in expanding their operations internationally—an area separate from China's Belt and Road Initiative for international investment, but one that will be equally important to watch as China's global clout grows, and pressures such as United States-China trade tensions persist.

Cross-Border Cultural Arbitrage

IPE's longstanding partnership with NRDC on both the PITI and the CITI could be considered cultural arbitrage, whereby IPE leverages synergies with NRDC to extend the influence of its work and raise its international reputation. The development of the PITI considered the state of environmental information disclosure in other countries, especially the United States and its TRI system—which provided significant inspiration to Ma Jun—and reports often highlight these cases as international best practices toward which China should strive. In fact, later PITI reports mention areas where Chinese environmental information disclosure standards are already in line with international best practices. In improving its own level of environmental information disclosure, China has also developed some systems that are unique and go beyond the standards in other countries, such as its extensive real-time disclosure of air emissions and wastewater data. One might argue that these systems came about because of factors unique to the Chinese context, such as difficulty in enforcing self-reporting.

With the CITI and IPE's green supply chain work, NRDC has helped to translate IPE's work into a narrative that Western actors can understand and value, especially senior management at major corporations. When corporations are first contacted regarding environmental violations, for international companies, letters are typically co-signed by NRDC along with IPE, making it more likely that the letter's recipient is familiar with the groups sending it. NRDC has also provided connections to staff at some brands in order to facilitate a path for IPE to contact them. In addition, all reports from the beginning have clarified that while the CITI is co-developed by IPE and NRDC, the scoring is implemented solely by IPE. In essence, the partnership is such that NRDC provides a platform for communication and legitimacy while IPE operates the tools and resources necessary to implement the joint "ask" of the two groups. Since IPE is capacity-constrained in both language and resources, partnering with NRDC helps to extend IPE's scope internationally while mitigating the scrutiny that Western business leaders might otherwise impose on a relatively small Chinese environmental NGO.

Discussion

IPE's stated mission focuses on restoring environmental quality to its fullest health. The organization aims to realize this goal by promoting information transparency and serving green development, objectives that correspond to IPE's two main standards, the PITI and CITI indices. Tracing the effectiveness of IPE's work from projects such as the PITI and the CITI to impacts can help us to discern what results derive from IPE's work. Using the commonly accepted evaluation framework of outputs, outcomes, and impacts, IPE's indices and other standardized processes can be taken as "outputs," with behavior change by governments, corporations, banks, and other advocacy targets representing "outcomes." Most difficult to measure, however, are "impacts," which entail concrete improvements in environmental quality or emissions reductions that can be causally traced to IPE's work.

A number of academics have attempted to conduct systematic evaluations of IPE's work. They find mixed results when investigating causal linkages between IPE's implementation of the PITI and actual impacts. Seligsohn et al. (2018) use the PITI to measure whether government transparency can substantively change environmental governance outcomes in the Chinese context. Transparency is proxied by looking at the degree of government information that is previously available to governments but newly supplied to the public, finding that transparency alone does not impact environmental quality outcomes and its impact remains constrained to the provision of information itself. It should be noted that Seligsohn et al. only use PITI data and corresponding environmental indicators from 2007 through 2011, so the analysis is constricted to the earliest years of the PITI. It is possible and even likely, therefore, that the study's findings would be different if conducted using data that also includes later years. However, Anderson et al. (2019) do find that IPE's implementation of the PITI increased local governments' compliance with environmental information disclosure regulations, thus reaffirming IPE's effect on outcomes with regard to government compliance. Putting this in the language of evaluation literature, IPE's work demonstrates outcomes, but impacts remain unclear and merit further exploration.

The difficulty in proving data-based causal linkages between IPE's work and environmental quality improvements is perhaps why IPE adopts outcomes as its main key performance indicators (KPIs) promoted on its website and used in reporting to its funders. For instance, as of July 2019, IPE's profile on the Skoll Foundation's website (Skoll 2019) touted the following:

- Working toward increased transparency, in 2017, over 160k factory violations were listed on IPE's public database, and 1,430 enterprises disclosed information about corrective actions undertaken.

- In 2017, 2,868 violation records were removed or solved from the infractions database, signaling that the documented issues were addressed.
- By the end of 2017, a total of 947 suppliers had completed and published 1,207 sets of Pollutant Release and Transfer Registry (PRTR) data. As such, a non-governmental PRTR system has begun to be piloted in China.

The above statistics all embody actions taken by targets of IPE's advocacy (e.g., enterprises disclosing information, removing violation records, and/or disclosing PRTR data) or, in the first case, actions taken by IPE itself (factory violations collected by IPE's public database). Underlying these outcomes is the more foundational outcome of greater information availability as a result of improved compliance with government environmental information disclosure regulations. What this means is that the outcomes reported to funders and others by IPE rely on the growing scope of the Blue Map Database, which results from increased government disclosure. With academic literature providing causal evidence between IPE's PITI, which could be considered an "output," and increased disclosure, these KPIs demonstrate the earlier-described relationship between IPE's simultaneous push for greater information availability via disclosure and its environmental compliance advocacy that relies on the data produced by the former. Put another way, these KPIs support Anderson's finding of the link between IPE's outputs and outcomes, but do not provide evidence to counter Seligsohn's results that did not establish a causal link between outcomes and impact. It thus remains unclear whether increased disclosure actually leads to improved environmental quality.

As for corresponding changes in China's institutionalized environmental governance processes, it could be argued that apart from merely increasing government compliance with existing regulations, IPE's standard-setting and Blue Map Database together provide motivation for the Chinese government to adopt even more progressive policies on environmental information disclosure where interests converge. Potentially the clearest example of such a hypothesized relationship is the previously discussed Measures on Self-Monitoring and Information Disclosure for Key State-Monitored Enterprises (Trial) following the Total Transparency Initiative led by IPE. While the TTI Initiative and adroit legal advocacy by working with NPC and CPPCC delegates certainly played a role in the policy's adoption, it could also be argued that without four years of PITI evidence indicating increased government disclosure and compliance with existing information disclosure regulations, the government may have been hesitant to adopt a more innovative policy that would require additional investments and capacity building.

Moreover, IPE's Blue Map Database and Green Choice Alliance network of over 50 grassroots Chinese environmental NGOs have also opened up space for civil society in China to leverage the resources developed through IPE's work. In IPE's green supply chain work, NGOs in the GCA are engaged for local-level pollution investigations. They also serve as a check on local protectionism by participating in the open review process of third-party factory audits. NGOs are involved in systematic efforts to regularly check factories' real-time data and make "micro-reports" on social media, as well as to report polluted bodies of water via IPE's "black and smelly rivers" (*hei chou he*) reporting platform, which submits data directly to the Ministry of Ecology and Environment and the Ministry of Housing and Urban-Rural Development to be used for rectifying and improving water quality. In sum, IPE's database and role as a convener massively extend the capacity of otherwise under-resourced groups by providing them a platform off of which they can base their activities.

While it is too early to say whether IPE's work may influence standard-setting for Chinese outbound investments, some multinationals have already incorporated IPE's tools into their own self-adopted standards (IPE and NRDC 2018, 2019). As one example, Inditex integrates checks against the IPE database into its supplier scorecards (Inditex 2014). A variety of other examples of how corporations have altered their processes as a result of the requirements of IPE's CITI index are discussed in more recent CITI reports.

To fully consider IPE's impact, one might also consider how stakeholders would perform if IPE and its actions were *not* a factor to be considered. On the government side, there may be no actor singularly tasked with enforcing environmental disclosure, whether voluntarily or otherwise. On the corporate side, a similar situation would exist for multinationals' supply chain environmental compliance in China. In sum, the *motivation* would be missing. But beyond just motivation, without the Blue Map Database, information would still be scattered across different sources and likely difficult to find. Self-motivated actors would be forced to directly engage the government, or create their own internal systems to replicate the function of the Blue Map Database. In short, absent IPE's roles as an intermediary and a regulator, a gap would likely exist between the Chinese government's actions to improve environmental enforcement and knowledge by multinationals and other corporations on how to leverage this information to strengthen their supply chain environmental management. Would any stakeholder act to fill this gap, and if so, who?

Given the premise of Guttman et al. that interactions between the state and nonstate actors in the context of environmental governance in China beg specialized consideration, where does IPE's unique advocacy model and theory of change stand? One major remaining question is whether the IPE model, consisting of the PITI and the CITI working together in tandem alongside the Blue Map Database, is replicable at scale outside of China, and in what context might it have positive impacts. What would the governance structures overseeing the data collection look like in different countries? In 2017, IPE started seeking to answer this question by collaborating with Business for Social Responsibility to conduct a feasibility study of the level of environmental information disclosure in Southeast Asia. This project established a starting point to begin adding some Vietnamese environmental violation records into the Blue Map Database by 2019; records from India have also been added. Will additional steps be taken with this information, and who will they impact? A second remaining question is how nonstate actors interact with the state in other non-Western countries, as the Guttman et al. article focuses on comparing the Chinese context with the so-called West. A better understanding of the ecosystem of stakeholders in developing countries might serve to improve IPE's future strategy toward potentially extending or replicating its model abroad.

Conclusion

This chapter discusses the role of IPE as a nonstate actor participating in China's environmental governance. It provides evidence to illustrate that IPE's influence extends beyond merely consolidating already-published government data. By engaging with the targets that IPE wishes to influence—such as local governments and corporations—in a consultative process, IPE effectively builds these actors' capacity for environmental compliance. IPE simultaneously employs both carrot and stick mechanisms: those who respond and engage improve both their assessment score and their governance capacity, while those who choose not to engage or improve expose themselves to repercussions through poor scores and low rankings. Beyond IPE, there exist additional negative pressures from central government assessments in the case of the PITI and market and consumer choice in the case of the CITI, whereby poor performance may lead to additional consequences. These incentivization mechanisms enable IPE's capacity to focus on working with those actors whose performance may lag but are willing and/or motivated to improve.

IPE's activities effectively extend the capacity of central-level environmental enforcement in China. By creating resources and mechanisms that actively supervise the compliance of local governments and polluting enterprises but also guide actors toward consistent compliance, IPE provides a service that complements government enforcement efforts. This is likely why authorities have evolved to tolerate and eventually support IPE's work. Nevertheless, IPE's causal effects are witnessed primarily in the form of outcomes, such as improved disclosure. Impacts are more difficult to measure because of the complexity of forces that lead to both improvements and backtracking in environmental quality. It is thus challenging to parse IPE's unique role among the multiplicity of factors at play, and this area merits further exploration. Still, the IPE model's positive effects on environmental governance outcomes pose lessons for other emerging economies grappling with issues like weak environmental enforcement capacity and local protectionism. Especially with Chinese overseas investment posing a risk to global environmental quality as well as the migration of industrial supply chains from China to other emerging economies, nonstate actors in other nations might learn from IPE's strategy of borrowing state authority to extend environmental enforcement capacity while positively rewarding consistent compliance.

Notes

1. The Blue Map Database can be accessed on IPE's website: http://wwwen.ipe.org.cn/index.html.
2. According to the Blue Map Database, "Major [data] sources include but are not limited to the following: The website of the Ministry of Ecology and Environment (MEE); 31 provincial environmental protection bureau (EPB) websites and provincial people's government websites; EPB websites and government websites of 388 cities; official websites of county-level environmental protection departments; environmental monitoring information from government supervisory departments as quoted by mainstream media; official Weibo accounts of government departments; official channels for the ministries of water resources, land and resources, housing and urban-rural development, and meteorology, among others" (IPE 2019).
3. "Key state-monitored enterprises" (*guojia zhongdian jiance qiye*) are enterprises selected for intensified supervision by China's national-level environmental ministry due to their high emissions. Research by IPE contends that they are collectively responsible for around 65% of China's industrial emissions (Stout 2013). The method for selecting which enterprises are included on the list is not publicly disclosed.
4. As a result of government reforms, China's MEP become the Ministry of Ecology and Environment ("MEE") in 2018. Its predecessor was the State Environmental Protection Administration, which was established in 1998 and persisted until MEP superseded it in 2008.
5. *Weilan ditu* in Chinese.
6. Referred to as "supervision records" (*jianguan jilu*) on both platforms, since not all records are expressly for violations.
7. There are some exceptions to data coming directly from government and other official sources; however, in all cases, IPE carefully checks and vets all data to ensure that it is in line with government regulations and policies. For instance, corporations may voluntarily disclose their annual emissions data based on a template provided by IPE that is based on disclosure regulations and largely collects data that has already been reported to the

government as part of routine inspections. In another example, during the "Take a Picture, Locate a Polluter" project, IPE checked all citizen reports and only uploaded those that matched with the official government-issued list of KSMEs.

8. The number of full-time staff fluctuates as positions are vacated and filled, but had grown consistently from 12 staff in 2013 to 35 staff at the end of 2017, to 38 staff as of June 2018.
9. In addition to the cities formally evaluated by IPE, a number of partner organizations including Green Anhui, Green Qilu, Fujian Green Home, Lvse Jiangnan, Jiangxi Environmental Communication Center, and Xingche Environment conduct regional assessments of additional cities in their respective jurisdictions.
10. These eight metrics include records of enterprise violations, results of enforcement campaigns against polluting facilities, clean production audit information, enterprise environmental performance ratings, disposition of verified petitions and complaints, environmental impact assessment (EIA) reports and project completion approvals, discharge fee data, and response to public information requests.
11. Following on the system established by the PITI, the maximum number of achievable points on indices developed by IPE is generally 100.
12. The main changes to the assessment criteria are detailed in the 2013–2014 PITI report, "Breakthroughs & Beginnings" (IPE and NRDC 2014).
13. A full explanation is provided in the 2013–2014 PITI report (IPE and NRDC 2014, p. 52): "Under this rule, a given evaluation's systematicness aspect score is used to limit the other aspect scores (timeliness, completeness, and user-friendliness), so that an assessment's final scores for the other three aspects are not allowed to exceed that assessment criteria's systematicness ranking. The reason for this rule is that the systematicness score measures the completeness, continuity, and regularity of published information, and primarily involves the quantity of published information. On the other hand, 'timeliness' and 'completeness' primarily involve the quality of the information, while 'user-friendliness' measures the quality of the publication itself. Because three of the aspects are assessed based on the published information, when scoring the last part we must increase the importance of the amount of information published compared to the amount which should have been published. The systematicness score includes a section on completeness, so it reflects to a greater extent the quantity of information published."

14. Since its inception, the CITI has generally used the Chinese *pinpai,* translated as "brand," to refer to its evaluation subjects. The CITI evaluation criteria 4.0, released in 2017, clarify this definition for the first time as follows: "An organization that conducts procurement from multiple levels of suppliers, coordinates production and distribution processes, and ultimately provides products to end users." This definition references the Supply Chain Risk Management Guidelines (GB/T 24420-2009). The term "brand" could be misleading because not all evaluation subjects are consumer-facing; for instance, Foxconn is an original equipment manufacturer (OEM) that generally supplies to major IT companies, such as Apple, rather than directly to consumers.
15. The GCA began as a coalition of 21 Chinese environmental NGOs that combined forces to launch the "green choice initiative" in March 2007 on World Water Day. This initiative, led by the Institute of Public and Environmental Affairs (IPE), originally aimed to leverage green procurement to promote environmentally responsible production as means of confronting the public health threat from industrial wastewater discharge. The GCA has since grown to include over 50 Chinese NGO members (IPE 2020).
16. "De-listing" functionally means that a company's record would be replaced with a sentence stating that the issue has already been addressed and accompanied by relevant information, such as a third-party audit report.
17. Enterprises may remove their environmental violation records from the Blue Map Database by passing a Green Choice Alliance (GCA) audit to prove that rectifications were successfully completed and emissions can consistently achieve compliance. GCA on-site audits are carried out by third-party auditing companies that have met a series of criteria. A representative of IPE or of an environmental NGO from the GCA also joins, reviews and monitors the audit process. Payments are negotiated between the enterprise and auditing company; IPE does not charge any fees for this process (IPE 2020).
18. A Pollutant Release and Transfer Register (PRTR) is an environmental database or inventory of potentially hazardous substances and/or pollutants released to air, water, and soil and transferred off-site for treatment or disposal (UNECE 2016). In August 2013, IPE established a voluntary disclosure platform for PRTR data in China. High environmental impact enterprises are encouraged to disclose their emissions and transfer data, particularly data regarding hazardous chemicals.
19. This brand map can be viewed here: http://wwwen.ipe.org.cn/MapBrand/Brand.html?q=6.

References

Anderson, Sarah E., Mark T. Buntaine, Mengdi Liu, and Bing Zhang. 2019. Non-Governmental Monitoring of Local Governments Increases Compliance with Central Mandates: A National-Scale Field Experiment in China. *American Journal of Political Science* 63 (3): 626–643. https://escholarship.org/content/qt6ss13791/qt6ss13791.pdf.

Cohen, Mark A., and V. Santhakumar. 2007. Information Disclosure as Environmental Regulation: A Theoretical Analysis. *Environmental & Resource Economics* 37 (3): 599–620.

Feng, Emily. 2018. Xiaomi Supplier Accused of Environmental Failings. *Financial Times*, May 27. https://www.ft.com/content/7dfeb4d6-5fef-11e8-9334-2218e7146b04.

Fortun, Kim. 2004. From Bhopal to the Informating of Environmentalism: Risk Communication in Historical Perspective. *Osiris* 19: 283–296. https://www.jstor.org/stable/pdf/3655245.pdf.

Guttman, Dan, Oran Young, Yijia Jing, Barbara Bramble, Maoliang Bu, Carmen Chen, Furst Kathinka, et al. 2018. Environmental Governance in China: Interactions Between the State and Nonstate Actors. *Journal of Environmental Management* 220: 126–135. https://www.sciencedirect.com/science/article/pii/S0301479718304924.

Inditex. 2014. Supporting Guide for Sustainable Manufacturing for Green to Wear Products. https://www.textilepact.net/pdf/publications/reports-and-award/supporting-guide-for-sustainable-manufacturing-for-green-to-wear-products.pdf.

IPE (Institute of Public and Environmental Affairs). Website. 2019. http://wwwen.ipe.org.cn.

IPE and Lvse Jiangnan. 2018. Ignoring Severe Supply Chain Pollution: Xiaomi Suspected of Violating IPO Disclosure Requirements. http://wwwen.ipe.org.cn/reports/report_19601.html.

IPE and NRDC. 2010. Breaking the Ice on Environmental Open Information: The 2008 Pollution Information Transparency Index (PITI). http://wwwen.ipe.org.cn/reports/report_1725.html.

IPE and NRDC. 2014. Breakthroughs & Beginnings: The 2013–2014 Pollution Information Transparency Index (PITI). http://wwwen.ipe.org.cn/reports/report_1762.html.

IPE and NRDC. 2018. Greening the Global Supply Chain—CITI Index 2018 Annual Evaluation Report Summary. http://wwwen.ipe.org.cn/reports/report_19695.html.

IPE and NRDC. 2019. Green Supply Chain 2019—CITI Evaluation Annual Report. http://wwwen.ipe.org.cn/reports/report_20163.html.

IPE and NRDC. 2020. Achievements of One Decade: 2018–2019 Annual Report of Pollution Information Transparency Index (PITI) for 120 Cities. http://wwwen.ipe.org.cn/reports/report_20260.html.

Kingdon, John W., and Eric Stano. 1984. *Agendas, Alternatives, and Public Policies*. Boston: Little & Brown.

Kostka, Genia, and Arthur P.J. Mol. 2013. Implementation and Participation in China's Local Environmental Politics: Challenges and Innovations. *Journal of Environmental Policy & Planning* 15 (1): 3–16. https://www.tandfonline.com/doi/full/10.1080/1523908X.2013.763629.

Lee, Hau, Erica Plambeck, and Maria Shao. 2009. Ma Jun and IPE: Using Information to Improve China's Environment. Technical report, Stanford Graduate School of Business, Teaching Case SI-115.

Li, Siman. 2016. Does Environmental Information Transparency Lead to More Collaborative Governance in China? An Analysis of the IPE Information Database's Function in Boundary Spanning. Lund University Master Thesis Series in Environmental Studies and Sustainability Science. http://lup.lub.lu.se/luur/download?func=downloadFile&recordOId=8878423&fileOId=8878430.

Lieberthal, Kenneth. 1997. China's Governing System and Its Impact on Environmental Policy Implementation. Woodrow Wilson International Center for Scholars, Environmental Change and Security Project: China Environment Series 1. https://www.wilsoncenter.org/sites/default/files/ACF4CF.PDF.

Ministry of Environment of the People's Republic of China (MEP). 2013. 国家重点监控企业自行监测及信息公开办法 (试行). http://www.gov.cn/gongbao/content/2013/content_2496407.htm.

Moustakerski, L. July. 2014. Publishing Pollution Data in China: Ma Jun and the Institute of Public and Environmental Affairs. http://ccnmtl.columbia.edu/projects/caseconsortium/casestudies/135/casestudy/www/layut/case_id_135.html.

NRDC China Program Website. 2020. https://www.nrdc.org/china.

O'Brien, Kevin J. 2013. Rightful Resistance Revisited. *Journal of Peasant Studies* 40 (6): 1051–1062. https://www.tandfonline.com/doi/full/10.1080/03066150.2013.821466.

Seligsohn, Deborah, Mengdi Liu, and Bing Zhang. 2018. The Sound of One Hand Clapping: Transparency without Accountability. *Environmental Politics* 27 (5): 804–829. https://www.tandfonline.com/doi/full/10.1080/09644016.2018.1452522.

Skoll. 2019. Institute of Public and Environmental Affairs Profile. Accessed July 8, 2019. https://skoll.org/organization/institute-of-public-and-environmental-affairs/.

Standaert, Michael. 2018. Chinese Smartphone Maker Draws Complaint Over Environmental Risk. *Bloomberg Law*, May 24. https://news.bloombergenvironment.com/environmentand-energy/chinese-smartphone-maker-draws-complaint-over-environmental-risk.

State Council of the People's Republic of China. 2007. 中华人民共和国政府信息公开条例. https://law.yale.edu/sites/default/files/documents/pdf/Intellectual_Life/CL-OGI-Regs-Chinese%281%29.pdf.

State Council of the People's Republic of China. 2017. 中共中央办公厅 国务院办公厅印《关于深化环境监测改革提高环境监测数据质量的意见》. http://www.gov.cn/zhengce/2017-09/21/content_5226683.htm.

State Environmental Protection Administration of the People's Republic of China (SEPA). 2007. 环境息公开办(试行). http://www.gov.cn/flfg/2007-04/20/content_589673.htm.

Stout, Kristi Lu. 2013. Can Social Media Clear Air over China? *CNN*, April 19. https://www.cnn.com/2013/04/19/world/asia/lu-stout-china-pollution/index.html.

Tarantino, Matteo. 2020. Navigating a Datascape: Challenges in Automating Environmental Data Disclosure in China. *Journal of Environmental Planning and Management* 63 (1): 67–86. https://www.tandfonline.com/doi/pdf/10.1080/09640568.2019.1659132.

Thaler, Richard H., and Cass R. Sunstein. 2009. *Nudge: Improving Decisions About Health, Wealth, and Happiness*. London: Penguin.

UNECE. 2016. About PRTR. https://prtr.unece.org/about-PRTR.

Wang, Alex L. 2018. Explaining Environmental Information Disclosure in China. *Ecology Law Quarterly* 44 (4): 865–924. https://papers.ssrn.com/sol3/papers.cfm?abstract_id=2956069.

Zhang, Lei, Arthur P.J. Mol, and Guizhen He. 2016. Transparency and Information Disclosure in China's Environmental Governance. *Current Opinion in Environmental Sustainability* 18: 17–24.

China Nonstate Actors and Global Environmental Governance

Water Stewardship: Engaging Business, Civil Society and Government in Collaborative Solutions to China's Freshwater Challenges

Michael Spencer and Zhenzhen Xu

"The highest good is like water".
Lao Tzu in the Dao Te Ching, 6th Century BC

The Alliance for Water Stewardship (AWS) is an international membership-based civil society organisation, with multi-stakeholder governance, offering a global standard for water stewardship, a verification system and a brand that allows water users to communicate responsible management of water resources. The AWS Standard, developed by a multi-stakeholder International Standard Development Committee,

M. Spencer (✉)
Monash Business School, Monash University, Melbourne, VIC, Australia
e-mail: michael.spencer@monash.edu

Z. Xu
Alliance for Water Stewardship Asia-Pacific, Melbourne, VIC, Australia
e-mail: zhenzhen@a4ws.org

D. Guttman et al. (eds.), *Non-state Actors in China and Global Environmental Governance*, Governing China in the 21st Century,
https://doi.org/10.1007/978-981-33-6594-0_7

is designed to address societal concerns about water; scarcity, pollution, ecosystem health, cultural use of water, equity of access to water, sanitation and health services related to water.

This case study will situate water stewardship within the literature on stewardship programs and review the evolution of water stewardship internationally and in China. It will consider the rapidly evolving environmental policy environment in China to contextualise AWS as an organisation seeking to engage industry in addressing China's severe water challenges. The study examines the interaction between state, industry and civil society in an area traditionally seen as a responsibility of government. It starts to build an understanding of the motives and constraints for industry participation in a non-state program and finds that AWS will need new collaborations and alliances, particularly with government agencies, to advance its mission of "igniting and nurturing leadership on credible water stewardship". To that end, it offers a preliminary model for future projects aimed at engaging clusters of businesses in becoming water stewards to achieve a scale that can address water pollution, scarcity and other challenges.

The case study is presented in four parts: Part one introduces AWS, its origins, methodology and activities in China; Part two establishes the context for water stewardship by describing the evolving policy environment; Part three examines the reaction of industry to water stewardship based on research with businesses in Kunshan and Tianjin; and, Part four presents a model for future development and learning that may be useful for other jurisdictions. The study reveals a number of obstacles for water stewardship but concludes that, despite these difficulties, a non-state actor such as AWS can contribute to China's desire to build an ecological civilisation.

Situating Water Stewardship

Soft Law, Certification and Stewardship

Stewardship programs can be characterised as systems having four elements: (1) a set of rules, generally referred to as a standard or requirements, that are intended to achieve environmental and/or social outcomes through participants complying with those rules; (2) a system for verifying compliance with those rules by organisations participating in the stewardship program; (3) a governance arrangement for managing the program and its rules, and; (4) a recognition system that rewards organisations complying with the program rules. The specific nature of stewardship programs will vary depending on the objectives, scope and theory of change, the stringency of the rules and verification processes and the nature of the governance arrangement.

A wide range of labels has been attached to these programs reflecting interest from different disciplines in their operation, drivers of success and impacts. These labels include; soft law, private governance, non-state market-driven normative systems, certification systems, standard-setting and voluntary environmental programs. Stewardship programs provide an alternative system of governance to traditional legal responses and sit within an area of research interested in normative or regulatory approaches that are voluntary and not dependent on the coercive powers of the state. Some are regarded as "greenwash" doing little or nothing to improve environmental performance. Others require participants to take action that they would not have taken in the absence of the program and thereby achieve superior environmental and/or social performance and positive externalities compared to non-participants (Cashore et al. 2004, 2007).

The impact of each program is generally dependent on the specific governance arrangements and the stringency of both the rules or standard with which companies are asked to comply and the stringency of the process for verifying compliance (Prakash and Potoski 2007). In many ways, these systems can be viewed as a contemporary application of Ostrom's design principles for managing common pool resources (Ostrom 1990; Ostrom et al. 1994).

Within this extensive literature, two relevant streams of research on stewardship programs are considered. One uses a broad definition of *voluntary environmental programs* and considers programs sponsored by non-government organisations, multilateral organisations such as the International Organisation for Standardisation (ISO), industry groups and government agencies (providing the programs are voluntary). This work sees stewardship programs as developing a form of club, after Buchanan's economic theory of clubs (1965), where participants (club members) receive benefits from an association with the "club's" positive brand in return contributing to broader social goals through positive externalities or addressing negative externalities. Brand association benefits from being a member of the club and accepting the club rules, include goodwill, reputation and social licence. By firms identifying as a *club member* stakeholders can differentiate between companies and bestow appreciation on those that are addressing externalities associated with their industry or region (Potoski and Prakash 2005, 2013; Prakash and Potoski 2007; Darnall et al. 2009).

While this work is relevant to an understanding of water stewardship, a second stream of work is relevant to understanding the contribution of non-state actors. This stream adopts a narrower definition of stewardship programs and considers non-government-organisations as drivers for market preference as well as contributors to multi-stakeholder forms of governance. These non-government organisations bestow a level of legitimacy on the program, its rules and the stringency with which those rules are enforced. Research in this stream has been based on stewardship programs in forestry, fisheries and agricultural commodities such as coffee (Cashore et al. 2004, 2007). Auld (2014) argues customer loyalty, market access or price premiums reward operators that help to address global environmental or social challenges.

Programs considered in this stream of work are often members of the ISEAL Alliance, an umbrella organisation for social and environmental standard setting and labelling organisations that provides guidelines for good practice and a form of meta-governance (Spencer 2008). AWS is a member of the ISEAL Alliance committed to multi-stakeholder governance and can draw on the learning from both of these streams of research.

Alliance for Water Stewardship

AWS was launched to develop and then implement an international system for defining, engaging and recognising good water stewardship by non-domestic water users; industry, agriculture, commerce, recreation and institutions. It built on several streams of work emerging in the first decade of the twenty-first century. The Millennium Drought in Australia saw many companies confronted by growing reputational, physical and regulatory risks from water scarcity. Business needed a credible framework for communicating water performance to stakeholders concerned about use of scarce water. This resulted in the establishment of Water Stewardship Australia (WSA) to develop a water stewardship system for Australia that would draw on the experience of other stewardship systems such as the Forest Stewardship Council and Marine Stewardship Council.

Just prior to this, the San Francisco-based Pacific Institute had highlighted the changing landscape of water risk facing corporations and that "almost without exception, corporations and investors are unfamiliar with freshwater-related risks" (Morrison and Gleik 2004). This led the Institute, in collaboration with Business for Social Responsibility (BSR), to publish a report in 2007 on developing and implementing a corporate water strategy to respond to the risks identified in the earlier report (Hwang et al. 2007). At this time, The Nature Conservancy freshwater team in the United States was exploring the potential for a freshwater ecosystem certification scheme to improve aquatic health particularly in the Great Lakes region. Shortly after, WWF US received funding for a water roundtable as the first step in a process to develop an international water stewardship standard. These strands came together in an international alliance, AWS, to develop a global water stewardship system in 2008 (Forbes 2018).

The centrepiece of the AWS system is the International Water Stewardship Standard (the Standard), supported by a credible verification and certification scheme, multi-stakeholder governance and a global water stewardship brand. The Standard was drafted by an International Standard Development Committee (ISDC) that included voices from business, civil society and government as well as different regions of the globe, developed and less developed countries and gender perspectives. Its task was to establish criteria and indicators for good water stewardship, against which companies could be certified by independent third-party auditors and gain access to the AWS brand.

AWS launched version 1 of the Standard in April 2014 at a meeting of the UN CEO Water Mandate in Lima Peru. The Standard sought to engage water-using sites (factories, farms, hospitals, etc.) in examining their catchment water context, site water context, and supply chain water use as well as stakeholder water concerns. Based on this analysis, it asked sites to consider their *shared catchment challenges* as well as site water risks and opportunities. This analysis provided the basis for developing a prioritised list of actions the site could take, both within its boundaries and in the broader catchment, to address the identified challenges, risks and opportunities. These actions were expressed as a *site water stewardship plan*. Sites were then engaged in a continuous improvement loop with the plan being implemented, evaluated and reviewed on an annual basis. To ensure transparency, the Standard required information on shared challenges and the *site water stewardship plan* to be communicated with stakeholders.

AWS recognised water stewardship was a journey and encouraged continuous improvement through a points system for achieving *advanced criteria* that enabled a water-using site to gain recognition as a core, gold or platinum water steward. To achieve these higher levels of recognition, in most cases required sites to engage in activities outside the factory gate, participate in collaborative projects or contribute to broader catchment goals. The AWS system (v.1) aimed to achieve four outcomes: sustainable water balance, good water quality, good water governance and healthy important water-related areas. These outcomes were reflected in most parts of the Standard but particularly in the implementation requirements where sites were generally required to work towards best practice (Abdel Al et al. 2014).

By the end of 2018, AWS had a multi-stakeholder, membership-based governance system, was a full member of the ISEAL Alliance, had an international network of credible third-party auditors and, listed 32 sites as certified to the AWS Standard including farms, food processing, chemicals and bottled water. There was evidence of an acceleration in adoption of the Standard with some 70 sites registered pending certification and Nestle Waters announcing plans to have all 85 of its water bottling facilities certified by 2025 (Alliance for Water Stewardship 2018). In 2019, AWS launched version 2 of the Standard, in compliance with an ISEAL requirement for regular review of member standards and added a fifth outcome on water and sanitation.

AWS in China

Rapid evolution of environmental governance in China over the past decade and a heightened focus on water created a strong interest in new ideas such as water stewardship following release of the AWS international standard. Chemical multinational Ecolab, whose financial support contributed to the development of the AWS Standard, offered its new plant at Taicang, north of Shanghai on the Yangtze River, as a pilot for testing the Standard and this became the first site certified to the Standard in 2015. This work generated interest from WWF China staff.

WWF Lower Yangtze Project Manager, Dr Aihui Yang, pursued this interest during a visit to Australia in 2014 where he reviewed projects underway in the Murray Darling Basin and the Mornington Peninsula near Melbourne. He also discussed the potential for collaboration with Water Stewardship Australia (WSA) to support the introduction of water stewardship into the WWF Lower Yangtze program. During a reciprocal visit, chapter co-author Spencer, then CEO of WSA, discussed opportunities in China with WWF and stakeholders from industry, government and academia. Based on this interest from China and other opportunities, WSA secured aid funding from the Australian Department of Foreign Affairs and Trade (DFAT) for an Indo-Pacific Water Stewardship Program and the appointment of chapter co-author Zhenzhen Xu as Regional Director based in Shanghai.

Collaboration between WWF China's Lower Yangtze program and the Alliance for Water Stewardship Asia-Pacific, into which WSA evolved, was formalised as the Taihu Basin Stewardship Network. The partners supported a social enterprise (*gongsi*) in Shanghai to develop the network's program by providing outreach, training and consulting services. Early collaboration with WWF's water stewardship pilots focused on the dairy industry and textile supply chain as both were under pressure over their water impacts. Textile production is said to be responsible for 20% of freshwater pollution in Asia (Sweden Textile Water Initiative 2015).

With most textile production in industrial parks, attention turned to how water stewardship could be adapted to work with groups of companies based in industrial parks. The majority of Chinese industry is located in national, provincial or municipal industrial parks. They offered scale, industry coherence and the prospect of working in collaboration with park management. This created the potential to prepare a single catchment analysis for multiple sites, engage a number of sites with similar water problems in training, sharing experiences and collaborating on solutions. However, the partners soon discovered the difficulty of engaging sites not familiar with voluntary "soft law" approaches and government officials more accustomed to engineering and regulatory solutions. There needed to be a more clearly defined and measurable set of benefits for firms to participate.

In 2016, AWS Asia-Pacific signed a Memorandum of Understanding with the TEDA EcoCentre in Tianjin to collaborate on developing and implementing a water stewardship program for industrial parks in Tianjin as well as nationally and internationally. TEDA EcoCentre is related to the Tianjin Environment Protection Bureau (EPB) and is based in the Tianjin Economic-Technological Development Area (TEDA). It is a lead organisation on environmental improvement for national level industrial parks in China. While the relationship was strong, the program failed to gain momentum. This was attributed to the region having a higher priority on air pollution while water issues associated with the industrial area were mainly being managed through large-scale government-funded infrastructure projects. However, a strong relationship was maintained and, over time, new opportunities emerged.

Building a Local Network

Building a strong network with government officials, academics and business was an important part of the development of AWS in China. For example, AWS Asia-Pacific co-sponsored an annual Taihu Forum, initiated by WWF in 2015, that brought together stakeholders in the heavily polluted Lake Tai region. Regional Director Xu regularly participated in Ecolab CEO water stewardship briefings with business leaders, academics and government officials. AWS commenced training programs for auditors, consultants, academics and non-government organisations. Conformity assessment bodies, audit firms, who could see the business opportunity of a water certification program, offered support in through access to training facilities, marketing to clients and sending their staff to training sessions.

Long-term connections with the China National Institute of Standardisation (CNIS) provided input to the AWS international Technical Committee and domestic legitimacy through the development of a China-specific version of the Standard. Dr Zhu Dongling, then with the Jiangsu provincial engineering consulting company, who had an interest in the potential of water stewardship to contribute to intractable problems in the Tai Lake basin, provided introductions to government bodies in Jiangsu province and created opportunities to test the water stewardship concept at his own tea farm. Xin Hao, CEO of Green Zhejiang, an environmental non-government organisation working on a range of environmental issues in the neighbouring Zhejiang province, and who had been a member of the AWS International Standard Development Committee, also provided introductions and an opportunity to trial application of the Standard at an agricultural site (Green Zhejiang's educational farm near Hangzhou).

Other non-government organisations such as the Society of Entrepreneurs for Ecology provided in-kind support and collaborated on a water education tool. In 2017, the first meeting of the China Water Stewardship Network convened to bring together stakeholders with an interest in the concept and its potential application in China. The network became a mechanism to share experiences and interest in water stewardship. These connections resulted in AWS receiving invitations to forums where water problems and solutions, particularly the engagement of industry, were discussed.

Building on Global Connections

The global connections AWS built outside China, through global supply chains and relationships established in customer countries, provided a starting point for engagement in three key water-using sectors; food and beverage, textiles and fashion and microelectronics. In food and beverage, relationships with multi-nationals such as Nestle and General Mills translated into domestic projects and demonstration sites that could sit alongside the Ecolab Taicang site.

In fashion and textiles, AWS Asia-Pacific established a collaboration with the Australian Fashion Council in 2015 that introduced Australian brands to their supply chain water risks. Brands were encouraged to look beyond their agents and CMT (cut, make, trim) factories to identify their textiles suppliers, dyeing and washing plants. Participating brands received a risk analysis of their textile suppliers that considered both catchment risk and individual site risks allowing brands to prioritise attention to suppliers that posed the greatest water risks. In one case, AWS provided training to in-house auditors from China and Bangladesh, to support them in promoting water stewardship. Moving to the next stage of suppliers adopting water stewardship will involve building greater collaboration between brands and linking the Australian program with other international fashion and textile environmental programs.

This industry-focused work strengthened collaboration with the China National Textile and Apparel Council (CNTAC) (see Chapter "The Governance Effect of Environmental CSR Reporting in China: State and Non-State Facilitation" for discussion of CNTAC) and allowed AWS Asia-Pacific to contribute to the development of that organisation's sustainable supply chain program. It also underpinned a relationship with the Better Cotton Initiative (BCI) resulting in collaborative projects with cotton farmers, coinciding with water stewardship principles being incorporated into the BCI standard and AWS providing training on water at BCI training sessions.

In microelectronics, the relationship started from contact between the AWS Regional Director and local representatives of major multinational brands and this was reinforced through global connections between AWS members or regional partners and the head office supply chain leads. This led to a substantial ongoing project with Apple to engage suppliers in water stewardship.

Situating Water Stewardship in China's Water Challenges

Water Challenges Facing China

Before proceeding to a deeper investigation of water stewardship in China, it is worth reflecting on the context and scope of the water challenges the country is facing. Water management was integral to the founding of the Chinese nation and its development over millennia. Post-1949 industrialisation changed China's relationship with water profoundly. In Hsiao Tung Fei's celebrated account of peasant life in China in the mid-1930s, he speculates the Tai Lake (Taihu) region of the Lower Yangtze had become the centre of silk production because; "according to the local people, [...] the good quality of the water" (Fei 1939, 16). By 2007, with 30 million people dependent on water from the lake, it was declared a natural disaster by the National Government due to algal blooms and bacteria pollution that left more than a million people in Wuxi City short of drinking water. The price of bottled water soared as the government froze new developments. Lake Tai, China's third largest freshwater reservoir and its most heavily polluted, became a symbol for the water challenges facing China (Duan et al. 2009).

Jiang (2018) says that China's post-revolutionary development had seriously over-stretched its natural resources and the country was facing one of the world's most severe water crises. China's renewable freshwater resources per capita is less than a third of the global average and only about one eight of North American availability (The World Bank 2019). This is compounded by the geographic distribution of water between and among regions, urbanisation, the impact of climate change on retreating glaciers as well as putrefaction and eutrophication of rivers and lakes. In 2011, two-thirds of cities suffered water shortages, more than 40% of rivers were severely polluted, 80% of lakes suffered eutrophication and about 300 million rural residents lacked access to safe drinking water (Liu and Yang 2012). Prime Minister Wen Jiabao said water shortages threaten "the very survival of the Chinese nation" (*The Economist* 2013). Hu and Tan (2018) claim that China, as well as India and Pakistan, "simply do not have sufficient water to ensure food and energy security plus develop under the current export-led economic growth model".

Addressing Water Challenges—Engineering Solutions

During China's long struggle with water management, mighty engineering projects that dam, divert, distribute and otherwise regulate the flow of water have been central to the nation-building efforts of successive dynasties. Recent history has reflected a similar emphasis. In January 2011, the government announced it would quadruple investment in water projects to four billion yuan (USD635 million) over the following 10 years. This would include upgrading 46,000 of the 87,000 dams and reservoirs built since 1950, construction of new dams, reservoirs and canals and undertaking the world's largest and longest water diversion project, the north-south water transfer project. Liu and Yang (2012) described the plan as laudable but not sufficient noting it focused largely on engineering works and failed to set targets for water quality. The Economist was more critical saying it represented a continuation of China's past fixation with engineering solutions and was likely to only provide a temporary reprieve: "By lubricating further water-intensive growth the current project may even end up exacerbating water stress in the north" (*The Economist* 2014). These solutions had a heavy cost with the Asian Development Bank estimating more than two per cent of China's GDP was spent trying to address its water issues, one of the highest levels in the Asian region (Asian Development Bank 2016).

Addressing Water Challenges—Regulatory Solutions

Increasingly, regulation has been strengthened as a tool to deal with China's serious water and air pollution problems. Environmental laws are not new to China, but the past 10 years has seen laws strengthened, enforcement tightened, and punishment increased. The Water Ten plan was released in 2015[1] and was followed by the Water Ten law in 2017[2] as well as a new environmental tax law the following year. The Water Ten plan, issued by the State Council, foreshadowed action to clean-up water quality, reduce the proportion of severely polluted water bodies and improve drinking water quality (State Council 2015). The Water Ten law provided for the suspension of new projects in regions that failed to achieve the discharge quota or water quality improvement targets. Penalties were increased ten-fold to one million RMB. The new tax law allowed provincial governments to impose water taxes based on local environmental conditions (Xu 2018).

While these regulatory changes were important, it was the stepping up of enforcement from 2016 that cut-through the cosy relationships between industry and provincial as well as municipal government. National environmental inspection groups began rolling two-year enforcement inspections in all provinces and cities resulting in factory closures, suspensions, charges, arrests and hundreds of imprisonments (including party officials). The crackdown left few in China, or in the global supply chains that had come to rely on China, in doubt that the government was serious about change. It was reinforced when national inspection groups revisited the provinces in 2018; 1494 people were held accountable in Yunnan; 162 people were detained in Guangdong, and; fines of 240 million yuan were issued in Jiangsu (Spencer and Ge 2019).

Government officials made it clear there will be no backing off the regulatory levers and this, as we shall see shortly, has been a driver for interest in AWS water stewardship. Gangie Li, Minister for Environmental Protection told media at the 19th Congress in 2017 that enforcement mechanisms were making "unprecedented achievements" and starting to transform pollution problems while raising public awareness of environmental protection. The promulgation of new laws and environmental inspections would continue (Xu 2017b).

New laws had not only given the Ministry of Water Resources and Ministry of Environmental Protection "more teeth" but the environmental tax law had brought-in the Ministry of Finance and engaged multiple government departments in addressing water scarcity and pollution (Tan 2018). This trend was consolidated in 2018 when the State Council undertook a major reshuffle of government administration with the creation of a new super-ministry of ecology and environment that consolidated environmental duties of six other departments with the former Ministry of Environmental Protection. Minister Li said the change recognised the need to overcome duplication in environmental management and separate policy or management functions from regulatory functions (Yang 2018).

As head of the new super-ministry, Li was well aware of the extent of the shift taking place, with a change of emphasis from economic development to sustainable growth, and the need to bring industry along. He said inspections were not contradictory to economic development even though there may be some impact on enterprises at a micro-scale. Change needed to be handled sensitively, he said, and the Ministry was opposed to a one-size-fits-all approach to enforcement. Li said violators should be given time to improve so that only those with no value or hope of improvement should be shut down (Xu 2017b).

Addressing Water Challenges—An Appetite for New Approaches

Importantly for AWS, the national drive to improve management of water and air stimulated an appetite for innovation. The new laws that accompanied the Water Ten Plan promoted market-based instruments such as price reform, tax and investment incentives and encouraged public participation and community supervision. This was also reflected in the national adoption of the River Chiefs system that applied pressure to provincial, municipal and town-level government and party officials to improve water management. The system had evolved out of the Tai Lake disaster in 2007 when Wuxi, the city seriously affected by the crisis, appointed officials at various levels of government as River Chiefs. When this was taken up at a national level in 2016, it required officials at four levels of government—provincial, urban, county and township—to adopt and take care of a certain stretch of river or lake.

Public supervision was introduced through a disclosure platform and officials had lifetime accountability for the environmental performance of their jurisdiction (Xu 2017a). By 2018, as some provinces extended the system to village level, the number of river chiefs was approaching one million (Huang and Xu 2019). This spreading of responsibility and accountability for water management, as well as the increased regulatory risks associated with water regulation, created necessary, although not sufficient, preconditions for AWS water stewardship.

Deepening Engagement with Industry

Breakthrough in Kunshan

Insofar as there was a breakthrough in the development of water stewardship in China, it came from work in the municipality of Kunshan near Shanghai and a partnership with the Kunshan Environment Protection Bureau (EPB). Jiangsu province, where Kunshan is located, has the largest direct industrial water use in China and the highest rate of embodied water use (Zhang et al. 2016). Dr Zhu Donglin, who had been part of the Taihu Basin Stewardship Network, had introduced the concept of water stewardship to the Kunshan EPB as they were dealing with impacts of the national government crackdown on water quality. The EPB had been forced to shut down major plants in the municipality for periods of time because of the water quality in local rivers and streams. While not all of the plants would necessarily survive the heightened focus on water and pollution, the municipality wanted to give polluters an opportunity to improve.

Initially, the EPB asked AWS and WWF China to work with major polluters in Qiandeng Town where there were compliance problems and past policies had not had any obvious impact. The project benefited from at least one senior staff member of the EPB being willing to look at alternative approaches as well as the concentration of advanced manufacturing firms linked to global supply chains. The AWS-WWF team developed a multi-stakeholder platform for water stewardship that engaged three universities as well as a conformity assessment body, TUV

Rheinland, and selected 12 enterprises to pilot and demonstrate the potential of the AWS Standard. This group was complemented by officials from relevant agencies in the municipality, including water and sewage departments, industry associations and grassroots environmental organisations. It was able to draw on technical support from the universities as well as expertise from AWS and WWF staff on water management improvement opportunities.

Over 12 months, the team provided training sessions, visits to water infrastructure sites as well as the Ecolab Taicang plant to understand both the opportunities and the business case for making improvements. After this pilot phase, the project was independently evaluated before being extended to the main industrial area, Kunshan Economic-Technological Development Zone (KETD) (Yang 2017).

Once it was established inside the KETD, the water stewardship project began to build credibility with major players in the industrial area. KETD covers an area of 115 km^2 and houses industries such as electronics, automotive components and logistics. The major water challenge facing the zone is pollution. Plant shutdowns as a result of pollution had caught the attention of global microelectronics brands. The program grew to involve 32 sites many of which were advanced printed circuit board (PCB) providers to these brands.

Apple already had a Clean Water Program operating within its supply chain and this very quickly became linked to the Kunshan water stewardship collaboration. AWS conducted training for Apple suppliers and worked with other microelectronics brands to engage their suppliers. Two of Apple's PCB suppliers in Kunshan became the first to embrace water stewardship and one became the second site in China to achieve gold level certification. These companies also became leaders within the zone working outside their fence-line to engage other stakeholders, such as schools, and peers in water stewardship. Apple became a member of AWS international and supported the regional office to scale-up to promote water stewardship to suppliers.

In 2018, Kunshan municipal government announced, as part of a series of measures to improve water, that it would offer an RMB 100,000 (USD 15,000) incentive payment to sites that successfully achieved AWS certification (Forbes 2018). This global first was significant for a number of reasons but particularly because it demonstrated a maturing relationship with government and the potential for AWS to work collaboratively to achieve shared water goals.

Deepening Understanding of Industry Attitudes to Water

With interest in AWS growing, there was a need to better understand how this could be leveraged beyond Kunshan. A research project with Monash Business School in Australia[3] set out to develop a more rigorous and systematic understanding of drivers and constraints affecting industry engagement with water stewardship. The project interviewed 15 sites in Kunshan and 14 sites in Tianjin using a detailed questionnaire. The interviewees included sites that were considering or had applied the water stewardship standard (16), sites not considering water stewardship (10) and sites that had not made up their mind (3). In Kunshan 12 of the sites that participated were in microelectronics while in Tianjin the sites were split between food and beverage (5), vehicles (3), electronics (2), chemicals (2) and waste processing (2). Sites in the two areas were of a similar average size, Tianjin 2,400 employees and Kunshan 2,735 employees, although there was quite a range of sizes in both areas. The two areas saw their water challenges somewhat differently with Kunshan sites giving a high priority to costs followed by pollution, industry impacts and the attitudes of management, workers and government authorities to water while counterparts in Tianjin were focused mainly on water efficiency and costs. This appears to suggest a greater consciousness of water impacts (e.g. pollution) in Kunshan where there was also a greater engagement with AWS.

As a first step, the research sought to understand current practice by sites and the gap they would be required to bridge to achieve AWS certification. Figure 1 illustrates current performance. The area for potential improvement is illustrated by the area above the bars and below the horizontal black line (28 sites completed these questions). The lower bars indicate areas with greatest potential improvement. These generally relate to issues outside of the factory gate such as examining catchment,

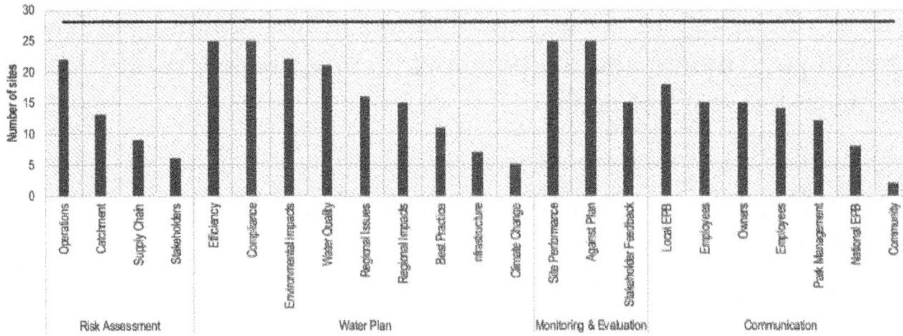

Fig. 1 Current practices of sites surveyed in Kunshan and Tianjin (Area between the horizontal line at the top showing potential alignment with water stewardship and the bars showing actual alignment represents are of potential improvement in water management by the group of sites surveyed)

supply chain and stakeholder issues in a water risk assessment, looking at climate change, infrastructure, best practice and regional issues in a water plan, asking for stakeholder feedback as part of a monitoring and evaluation process and communicating more broadly on water performance. The sites surveyed were generally good water managers suggesting the scope for improvement with weaker performers would be considerably greater. There was a weak association between current performance and interest in water stewardship. Literature on stewardship programs suggests the strongest interest will come from those just below the very best performers with less interest from the worst performers.

Understanding Motivations and Constraints

At the outset of this research, it was assumed that decisions to participate or not participate in water stewardship would be based on a rational analysis of costs and benefits. What emerged was that, while there was a relationship between a site's perceived costs and benefits and a site considering water stewardship, it was not a strong relationship. In other words, a positive perception of net benefits from water stewardship did not mean a site would consider water stewardship and a perception of negative net benefits did not mean a site would not consider water stewardship. This is illustrated in Fig. 2 where each site's rating of costs and

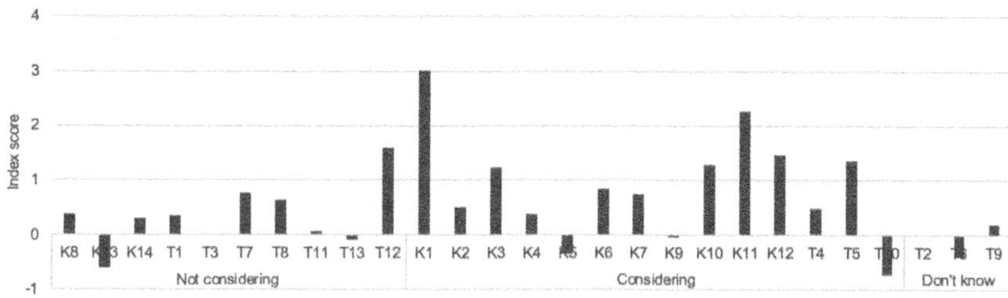

Fig. 2 Perceived net benefits for sites considering and not considering water stewardship (K = Kunshan, T = Tianjin)

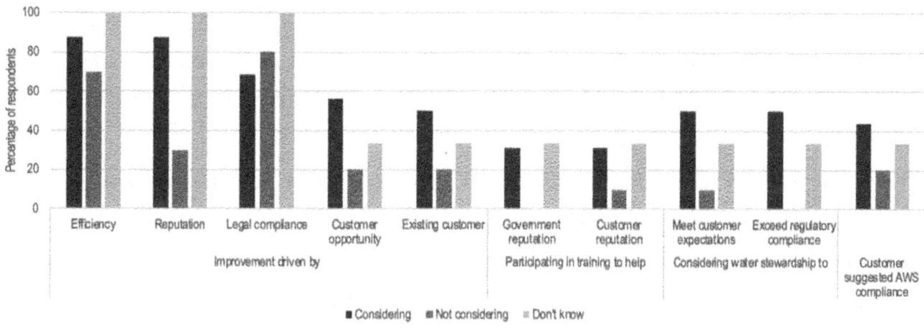

Fig. 3 Selected site responses to drivers of existing improvement, participation in water stewardship training, reasons for considering training and, customer suggestions to comply with AWS water stewardship

benefits has been used to calculate a net benefit index score. There is at best a weak correlation between positive net benefits and decisions to consider water stewardship or the opposite. It emerged that the decision-making process for sites was more complex and involved consideration of a range of issues beyond *rational choice*.

Western literature on stewardship systems has focused on reputation (Potoski and Prakash 2005, 2013; Prakash and Potoski 2007; Darnall et al. 2009), customers and markets (Cashore et al. 2004, 2007; Auld 2014) as drivers for participation. However, in China this research has indicated that the influence of government and the reputation of the firm with government also needs to be considered. In Fig. 3, the importance of government is observed from a variety of perspectives. When asked what had driven site water improvements already made or planned, legal compliance was second only to efficiency improvement. Reputation was important, but only for sites already considering water stewardship. Asked about their participation in water stewardship training, enhancement of their reputation with government was as important as reputation with customers. And, when asked why they were considering participating in the water stewardship system, exceeding regulatory requirements was as important as meeting customer expectations. Of 10 sites whose customer had suggested they comply with water stewardship requirements, only seven were considering participation, two were not considering and one was undecided. Two observations emerge; (1) customer pressure is important but does not guarantee participation and, (2) a strong driver for site engagement with water stewardship appears to be regulatory pressure, a desire by sites to get ahead of regulators and reputation with government.

Responses on the importance of government, customers and reputation differed between sites considering and not considering water stewardship. Those considering water stewardship were more aware of customer concerns about water and more likely to see market opportunities and reputational benefits from engaging with water stewardship. Anecdotally, it was noted that these companies tended to have stronger links with international markets through ownership or customer relationships. On the other hand, sites not interested in water stewardship, placed a greater emphasis on compliance with government regulations as a driver for making water improvements.

The emphasis on government was also evident when sites were asked how they see responsibility for water and catchment health spread between different actors. Sites favourable to water stewardship identified a greater spread of responsibility between catchment stakeholders whereas sites not interested in water stewardship placed a greater emphasis on government as the key actor responsible for water and catchment health. There are likely to be two factors at work here, both the role of the state in relation to water and, a theme developed in some Western literature arguing the good work of water engineers in dealing with water challenges had made water invisible to customers and other stakeholders (e.g. Sharp 2017).

This leads to two further observations: (1) water stewardship will face cultural barriers where water is seen primarily as a government responsibility and sites see themselves as having only a relatively small responsibility for water and catchment health, and; (2) government support for new water stewardship programs will be important to establish both the legitimacy of the program and overcome these cultural barriers. Industry will need to be persuaded that it has an important role in addressing water and catchment challenges. To engage industry, particularly during the early stage of developing a water stewardship program, may require *borrowing authority* from a relevant government agency (Guttman et al. 2018; Green and Auld 2017).

There was a strong expectation by respondents that water stewardship would deliver operational benefits (improved efficiency, lower costs, etc.). This was surprising given sites could access consultants to provide this advice. Researchers such as Prakash and Potoski (2007) dismiss private operational benefits as having no analytic value to understanding *voluntary environmental programs* because, they argue, a profit oriented firm would take action that delivered private benefits regardless of whether they were a member of the reputational *club* or not. The contrary view, articulated by Porter and van der Linde (1995), is that firms do not necessarily have perfect knowledge of opportunities for innovation and that existing organisational incentives may not align with adopting innovative opportunities.

It was reported anecdotally in Guttman et al. (2018) and confirmed in this research that sites in China were seeking technical advice from the water stewardship program on how and what sort of operational improvements they could make. Operational benefits were the most sought-after benefit from participation in the water stewardship program. This indicates that water stewardship either provided an additional incentive to seek this advice or was an opportunity to gain an understanding of improvement opportunities from a neutral party. Either way, this would need to part of any future program.

Building a Model for Future Development of Aws Stewardship

Taking Water Stewardship Forward in China

The research points to three observation for taking water stewardship forward; (1) in addition to customer pressure, in China there is a need for government pressure for change, (2) there is a need to offer operational improvement advice and, (3) there is a need to overcome attitudinal obstacles to the role of industry in water management and catchment health. Observations (1) and (3) present a problem because the support of even sympathetic government officials cannot be guaranteed. As one municipal official commented after a presentation on water stewardship: "You have presented a very interesting idea to us on bottom-up engagement on water issues but as you know, China works through top down regulation, so I am not sure about the role for water stewardship in China". AWS will need to build trust, engagement and support from local officials while demonstrating the capability of water stewardship to create shared benefits and positive impacts.

In discussions with academics and government officials in China, the idea that the Kunshan experience may have been a "one-off" and was not likely to be repeated was canvassed. Responses included variations on a theme that China was not a homogeneous whole on environmental issues and that while the national priority on environmental improvement would remain a mainstay of the political environment, how it was acted upon would vary among provinces, counties and even townships. It would depend on how officials in each area were working through the prioritisation of economic and environmental issues and then within the palette of environmental challenges, the priority attached to water. Even where water was a priority, the performance framework for local officials would be a consideration. For example, was the implementation of an engineering plan or was regulatory enforcement strategy the main priority? It was noted that engineering and regulatory approaches may have more immediate impacts, offering stronger recognition for local officials, than a management reform program such as water stewardship that might take time to deliver measurable impacts. That is not to say water stewardship would not find a place rather, it needs to select target areas carefully based on these considerations.

In this regard, it was suggested that an area such as the Lower Yangtze, would provide a more fertile and receptive field for water stewardship than Tianjin in northern China given the national government has nominated water quality as a priority for this region. Where the pressure is greatest local officials might be more open to collaborative solutions, particularly where there is support from an organisation such as TEDA EcoCentre, a national lead agency on environmental issues for industrial parks, and water stewardship can use its track record from Kunshan (particularly if this experience can be positioned as a national pilot). The program would still need to find local champions who were respected by government and the party and, perhaps, less constrained by rigid performance criteria. Strong support from municipal officials had been an important factor contributing to success in Kunshan. The program would need support from officials who, as Minister Li had suggested, were willing to support industry through an improvement process. It was felt this combination of factors would be possible with careful discussion and patience.

Building Collaboration Based on Measurable Impacts

For AWS to build relationships with local agencies, there will need to be a shared interest in promoting industry engagement in water management and a champion within a relevant agency. In addition, to be sustainable over the longer term, AWS will need to be able to demonstrate positive impacts on water and catchment health in terms that are meaningful. While it will take time to produce measurable impacts, it is hypothesised that this combination can create potential for the *diffusion of public authority* through a non-state actor such as AWS. As a first step towards building a framework for measuring and reporting catchment and water impacts, AWS Asia-Pacific investigated the potential of using the System of Environmental-Economic Accounts (SEEA). The SEEA framework provides a common language to measure and track environmental impacts (including water quality and quantity) and catchment health. SEEA, designed to sit alongside the System of National Accounts, was developed through a collaboration between national governments and multilateral organisations coordinated by the United Nations Statistical Office.

The AWS standard contains a number of data points for tracking change over time as a result of industry participation. These include measures at both a site and catchment level related to; water balance, water quality, important water-related areas and catchment health. IDEEA Group (2018) was engaged to examine the potential for the SEEA, in conjunction with the AWS Standard, to provide a reporting framework in Tianjin and Kunshan. IDEEA found that SEEA could help to communicate water stewardship performance to central and provincial government and frame local data held at a municipal or provincial level. This would strengthen the rationale for government to collaborate with AWS and make environmental investment decisions based on impacts that could be associated with water stewardship participation. It would also create a compelling monitoring and evaluation tool to ensure AWS was delivering promised outcomes.

Building on the potential, over time, to report measurable impacts, the Monash survey tested various forms of assistance that might be offered to encourage participation; incentive payments, reduced taxes, levies and charges, discounted loans (green capital) and, reduced compliance burden for sites that are compliant with the water stewardship standard. It was anticipated these might improve the net benefit equation for sites considering water stewardship by reducing compliance costs or costs associated with new equipment to improve water management. Generally, sites were positive about any form assistance even though this would not necessarily be the difference between participation and non-participation.

Sites in Kunshan where incentives had already been offered were positive about incentive payments but were also supportive of measures that would reduce their compliance burden. Some sites had said that the reporting burden to government was already quite onerous. In Tianjin, where a majority of sites were not considering water stewardship and the major challenge was seen as water costs, sites were more positive about reductions in taxes, levies and water charges. Green financing to support investment in improved infrastructure was favoured less. Thus, while it is important to remember that decisions to embrace water stewardship are not based purely on a rational cost-benefit decision, financial support may contribute to encouraging participation by improving the business case and enhancing legitimacy through government support. The potential to report measurable impacts would strengthen the attraction of water stewardship as a public policy tool.

A Model for Future Collaboration

Experience in China suggests a more nuanced model for building participation in water stewardship than is generally found in the academic literature on stewardship programs. The Monash research has tentatively identified a hierarchy of success factors for what has been referred to as cluster-based water stewardship programs (Fig. 4). The base of the pyramid illustrates the need for legitimacy that is created through support from key stakeholders and, in particular government stakeholders who can help to address the cultural barriers to industry involvement in water and catchment health. Legitimacy will also be

Fig. 4 Hierarchy of success factors for cluster-based water stewardship programs in China

influenced by factors such as the priority on water in the region, immediate threats from poor water management, the level of international exposure, involvement of customers, the availability of early adopters and attitudes of key local officials. The next level identifies the need to address business case issues such as creating opportunities for operational improvement through the water stewardship program. Incentives and other investments from government, customers and philanthropists can assist at this level. The third level highlights concern about the compliance burden industries face and the risk to business continuity from regulatory enforcement. The final level deals with opportunities for reputational benefit for the site from water stewardship.

While much of the literature on stewardship programs has focused on reputational benefits, this research finds that reputation is a consideration once the other levels of the pyramid have been engaged (hence the survey found that reputation was only a strong consideration for sites already committed to water stewardship). What should be noted, however, is that this prioritisation may change over time. In other words, as the brand of water stewardship becomes better known (greater *brand value*), this may address both legitimacy concerns and enhance the business case. This would certainly be consistent with findings in the literature (Potoski and Prakash 2013).

Building on this hierarchy, a preliminary model for implementation of cluster-based AWS programs was developed. Exactly how interactions play-out will depend on the particular circumstances of the region, municipality or catchment where the program is to be developed. Figure 5 illustrates the phased approach to developing a water stewardship project by first establishing that the preconditions exist, identifying key stakeholders and building alignment with those stakeholders. At a minimum it would be expected in China that the key stakeholders would include a local regulatory agency or water leader (Environmental Protection Bureau, River Chief). Others bringing important relationships to the project would be early adopters, customers and non-government organisations.

The preparatory stage would involve understanding catchment challenges (as required in the AWS Standard) and potential responses to those challenges as well as building a multi-stakeholder collaborative team.

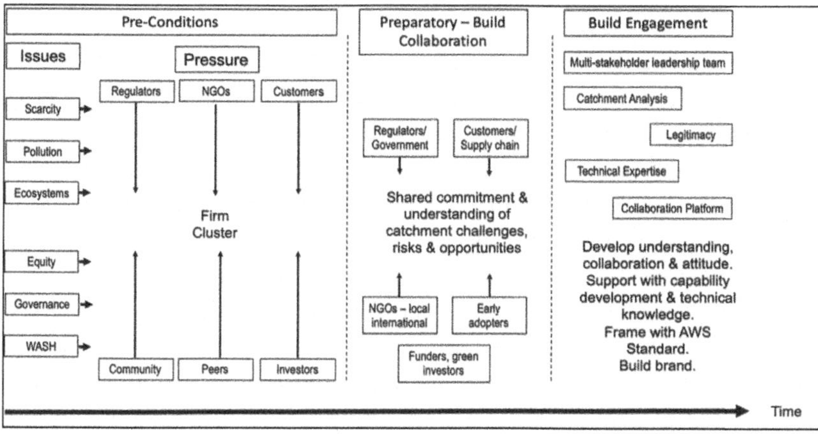

Fig. 5 Provisional model for building a cluster-based water stewardship program

Engagement would be built by this team (and the legitimacy it provides) through the catchment analysis as well as external technical expertise and knowledge sharing through the project team and among participating sites. This approach would help build the brand of water stewardship in the cluster.

As a cluster project develops, tracking of impacts will be an important part of monitoring and evaluation to reinforce credibility and legitimacy and gain potential investment from agencies, philanthropists, customers or financiers interested in the public benefits water stewardship can deliver. These would build a virtuous cycle of investment and participation in addressing shared water challenges. It is envisaged this model could be applied to both industrial areas (industrial parks) and agricultural areas (e.g. a village water stewardship program).

Learning from China for Global Programs

Whether the model described above is transferrable beyond China remains to be seen and will require outcomes from additional research currently underway in other jurisdictions. However, we should avoid assuming that the potential of water stewardship in China is dependent solely on the political circumstances. The experience of AWS in China would suggest it has as much to do with the point in time China has arrived at where the desire for economic development is now facing a counterbalance from a desire for a healthy and sustainable environment. As such this experience and the model developed above could be relevant to other countries arriving at a similar point in their development. A further consideration is the level of water stress China is currently facing with both scarcity and water quality being major constraints on continuing development and a significant cost borne by Chinese society. In that regard, the model for cluster development could provide elements of an approach in other countries facing similar issues of water stress where there is a will by government agencies to engage business (and by business to work with government) in addressing water challenges and the causes of water stress.

Conclusions

China is at a turning point in environmental governance as it seeks to work though the theory and practice of building an ecological civilisation. How the COVID-19 pandemic will affect this remains to be seen. However, prior to the pandemic there was a desire to find a different pathway to that followed by Western countries. It will take time to change attitudes and practices associated with water that have developed over millennia and result in a default assumption that water challenges can be resolved solely by engineering and regulatory responses. There will be setbacks as new priorities emerge (particularly in the current challenging times) and the ways of the past seek to re-assert themselves but, there is undoubtedly strong commitment from the current leadership to change. Water stewardship's appearance in China during this time creates exciting opportunities for multi-stakeholder approaches to environmental stewardship, to promote the active involvement of all in improving environmental management.

While the reception of water stewardship has been positive, exactly how this plays out and how the new approach to water finds a place is a work in progress. What is clear is that traditional Western approaches continue to build unresolved conflicts; between economic benefit and environmental management, between private interests and public interests and between the needs of the present generation and the rights of future generations. In at least searching for alternative pathways, China has created interesting preconditions for future growth of water stewardship that offers new opportunities for collaboration between state and non-state actors. Win-win outcomes on pollution control and management of environmental externalities would seem to be an important ingredient for building the next stage of China's evolution. Water stewardship, with its strong foundation in Ostrom's work on managing common pool resources, can offer such outcomes and contribute to building an ecological civilisation in China.

Notes

1. The *Water Pollution Control Action Plan* (2015). See: http://www.mee.gov.cn/zcwj/gwywj/201811/t20181129_676575.shtml.
2. The *Water Pollution Prevention Law* was approved by the Standing Committee of the National People's Congress on the 27th of June 2017 to become operational from the 1st of January 2018. See: http://www.mee.gov.cn/ywdt/hjywnews/201706/t20170628_416809.shtm.
3. Including some financial support from the Australian Water Partnership (an Australian Aid program).

References

Abdel Al, Imane, Shahid Ahmad, Maureen Ballestero, Sanjib Bezbaroa, Peter Cookey, Axel Dourojeanni, Carlo Galli, et al. 2014. The AWS International Standard. Alliance for Water Stewardship.

Alliance for Water Stewardship. 2018. AWS Certification. Accessed 18 September 2018. http://a4ws.org/certification/.

Asian Development Bank. 2016. *Asian Water Development Outlook 2016: Strengthening Water Security in Asia and the Pacific*. Mandaluyong City, Philippines: Asian Development Bank.

Auld, Graeme. 2014. *Constructing Private Governance: The Rise and Evolution of Forest, Coffee and Fisheries Certification*: New Haven: Yale University Press.

Buchanan, James M. 1965. An Economic Theory of Clubs. *Economica* 32 (125): 1–14.

Cashore, Benjamin, Graeme Auld, and Deanna Newsom. 2004. *Governing Through Markets: Forest Certification and the Emergence of Non-State Authority*. New Haven: Yale University Press.

Cashore, Benjamin, Elizabeth Egan, Graeme Auld, and Deanna Newsome. 2007. Revising Theories of Nonstate Market-Driven (NSMD) Governance: Lessons from the Finnish Forest Certification Experience. *Global Environmental Politics* 7 (1): 1–44.

Darnall, Nicole, Matthew Potoski, and Aseem Prakash. 2009. Sponsorship Matters: Assessing Business Participation in Government- and Industry-Sponsored Voluntary Environmental Programs. *Journal of Public Administration Research and Theory* 20: 283–307.

Duan, Hongtao, Ronghua Ma, Xiaofeng Xu, Fanxiang Kong, Shouxuan Zhang, Weijuan Kong, Jingyan Hao, and Linlin Shang. 2009. Two-Decade Reconstruction of Algal Blooms in China's Lake Taihu. *Environmental Science and Technology* 43 (10): 3522–3528.

Fei, Hsiao Tung. 1939. *Peasant Life in China: A Field Study of Country Life in the Yangtze Valley*. London: Kegan Paul, Trench, Trubner & Co., Ltd.

Forbes, Mark. 2018. *The Evolution of Water Stewardship: An Australian Perspective*. Canberra.

Green, Jessica F., and Graeme Auld. 2017. Unbundling the Regime Complex: The Effects of Private Authority. *Transnational Journal of Environmental Law* 6 (2): 259–284.

Guttman, Dan, Oran Young, Yijia Jing, Barbara Bramble, Bu Maoliang, Carmen Chen, Kathinka Furst, et al. 2018. Environmental Governance in China: Interactions Between State and 'Nonstate Actors'. *Journal of Environmental Management* 220: 126–135.

Hu, Feng, and Debra Tan. 2018. No Water, No Growth: Does Asia Have Enough Water to Develop? *China Water Risk*. Hong Kong.

Huang, Qidong, and Jianjun Xu. 2019. Rethinking Environmental Bureaucracies in River Chiefs System (RCS) in China: A Critical Literature Review. *Sustainability* 11 (6): 1–13.

Hwang, Linda, Sissel Waage, Emma Stewart, Jason Morrison, Peter H. Gleick, and Mari Morikawa. 2007. At the Crest of a Wave: A Proactive Approach to Corporate Water Strategy. Business for Social Responsibility, Pacific Institute.

Institute for Development of Environmental-Economic Accounting. 2018. Application of AWS and SEEA to Accounting for Water: China Industrial Park Demonstration (Final Report). Melbourne.

Jiang, Min. 2018. *Towards Tradable Water Rights: Water Law and Policy Reform in China.* Berlin: Springer.

Liu, Jianguo, and Wu Yang. 2012. Water Sustainability for China and Beyond. *Science* 337 (6095): 649–650.

Morrison, Jason, and Peter Gleik. 2004. *Freshwater Resources: Managing the Risks Facing the Private Sector*. Oakland, CA: Pacific Institute.

Ostrom, Elinor. 1990. *Governing the Commons*. New York: University of Cambridge Press.

Ostrom, Elinor, Roy Gardner, and James Walker. 1994. *Rules, Games and Common Pool Resources*. Ann Arbour: The University of Michigan Press.

Porter, Michael E., and Claas Van der Linde. 1995. Towards a New Conception of the Environment-Competitiveness Relationship. *Journal of Economic Perspectives* 9 (4): 97–118.

Potoski, Matthew, and Aseem Prakash. 2005. Green Clubs and Voluntary Governance: ISO 14001 and Firms' Regulatory Compliance. *American Journal of Political Science* 49 (2): 235–248.

Potoski, Matthew, and Aseem Prakash. 2013. Green Clubs: Collective Action and Voluntary Environmental Programs. *Annual Review of Political Science* 16: 399–419.

Prakash, Aseem, and Matthew Potoski. 2007. Collective Action Through Voluntary Environmental Programs: A Club Theory Perspective. *The Policy Studies Journal* 35 (4): 773–792.

Sharp, Liz. 2017. *Reconnecting People and Water: Public Engagement and Sustainable Urban Water Management, Earthscan Water Text Series*. London, UK: Routledge.

Spencer, Michael. 2008. Governance Options Paper. Third Water Stewardship Stakeholder Forum, Melbourne, October.

Spencer, Michael, and Sally Ge. 2019. Water Stewardship in the Supply Chain. Legacy responsible fashion summit, Sydney, March 14.

State Council. 2015. China Announces Action Plan to Tackle Water Pollution. The State Council of the People's Republic of China. Accessed

2 April 2019. http://english.gov.cn/policies/latest_releases/2015/04/16/content_281475090170164.htm.

Sweden Textile Water Initiative. 2015. Global Results 2015; Collaborating to Drive Global Change Towards Sustainable Textile and Leather Production. Stockhom International Water Institute.

Tan, Debra. 2018. 5 Trends for 2018: The Year of the Dog. *China Water Risk*.

The Economist. 2013. Water in China: Desperate Measures. *The Economist* 409 (8857), October 12.

The Economist. 2014. Water Consumption: A Canal Too Far. *The Economist* 412 (8906): 44–45.

The World Bank. 2019. World Development Indicators. Accessed 13 March 2019. https://datacatalog.worldbank.org/dataset/world-development-indicators.

Xu, Yuanchao. 2017a. China's River Chiefs: Who Are They? *China Water Risk*. http://www.chinawaterrisk.org/resources/analysis-reviews/chinas-river-chiefs-who-are-they/.

Xu, Yuanchao. 2017b. Green Development for a Beautiful China. *China Water Risk*.

Xu, Yuanchao. 2018. 5 Laws to Watch Out For In 2018. *China Water Risk*.

Yang, Aihui. 2017. Industrial Park Water Stewardship in China: the Case of Kunshan. Alliance for Water Stewardship Global Forum, Edinburgh, UK.

Yang, Wanli. 2018. New Ecological Environment Ministry Is a Milestone. *China Daily*, March 17. http://www.chinadaily.com.cn.

Zhang, Bo, Zhanming Chen, Li Zeng, Han Qiao, and Bin Chen. 2016. Demand-driven Water Withdrawels by Chinese Industry: A Multi-regional Input-output Analysis. *Frontiers of Earth Science* 10 (1): 13–28.

Non-industry and Nonstate Actors Contribution in the Standard Drafting Process: Examples from the Development of China Room Air Conditioner Standards

Xiaopu Sun and Richard ("Tad") Ferris

Introduction

Nonstate actors play various roles in the processes of national policy-making and implementation in China and around the world. In this chapter, the authors endeavor to present key strategic insights into how non-industry and nonstate actors, including foreign research institutions and subject matter experts, can get involved in and maximize their contributions to China's national standard drafting process. They do so

X. Sun (✉) · R. Ferris
Institute for Governance & Sustainable Development (IGSD), Washington, DC, USA
e-mail: xsun@igsd.org

Institute for Governance & Sustainable Development (IGSD), Paris, France

R. Ferris
e-mail: tferris@igsd.org

D. Guttman et al. (eds.), *Non-state Actors in China and Global Environmental Governance*, Governing China in the 21st Century,
https://doi.org/10.1007/978-981-33-6594-0_8

through the "lens" of China's development of national, compulsory air conditioner (AC) standards.[1]

In Section I, the authors set the stage for discussion by describing the importance of cooling efficiency for the world and in particular for China. More specifically, the authors reflect on China's leadership role in the development of the Montreal Protocol on Substances That Deplete the Layer (Montreal Protocol) and the associated Kigali Amendment. They also explain the significant climate mitigation potential from the duel strategy of phasing down HFCs and improving energy efficiency of ACs. To facilitate reader understanding of national standard drafting in China, the authors also describe fundamental aspects of China's national regulatory system governing AC energy efficiency, including the compulsory energy efficiency standards.[2]

In Section II, the authors then provide their insights on key strategy recommendations for non-industry and nonstate actors to maximize their contribution in raising the climate and energy efficiency ambition of national standards. In so doing, the authors provide illustrative examples drawn from China's recent national AC energy efficiency standard drafting process.

Setting the Stage for Maximizing the Contribution of Non-industry and nonstate actors in the Standard Drafting Process

Cooling Efficiency's Importance for China

The warming climate is significantly elevating global cooling demand. Research calculates that over 1.1 billion people globally face cooling access risks, threatening the ability to escape poverty, to keep their children healthy, vaccines stable, food nutritious, and economies productive (SEforALL and K-CEP 2019). In a warming world, global AC use is growing fast, driven by demand in emerging economies with hot climates and rising incomes that are undergoing rapid urbanization and electrification.

Cooling equipment is critical to management of the world's energy demand and climate, with co-effects on air pollution, food security, and public health. Global sales of ACs have been growing steadily and significantly.[3] Globally, the number of cooling appliances in use is estimated to grow from 3.6 billion today to 9.5 billion by 2050 (UNEP 2019). According to the International Energy Agency (IEA), global energy demand for ACs is projected to triple by 2050, and when met with the current fossil fuel-heavy electricity generation will nearly double greenhouse gas (GHG) emissions from this sector, from 1.25 billion tonnes in 2016 to 2.28 billion tonnes a year in 2050, further increasing the world's need for cooling in a dangerous feedback loop (IEA 2018).

China leads the global AC market by producing about 70% of the world's room ACs (IEA 2019). Domestically, China's energy demand for spacing cooling in buildings has been increasing at 13% per year since 2000, which accounted for about 16% of peak electricity load in 2017 (IEA 2019). In hot summer days, the share spurs to as much as 50% of peak electricity demand (IEA 2019). Due to the coal-combustion dominated power generation in China, cooling-related carbon dioxide (CO_2) emissions from electricity consumption had increased fivefold between 2000 and 2017, in addition to other related GHG emissions such as refrigerants including hydrofluorocarbons (HFCs) and black carbon (IEA 2019). It is no surprise, then, that China's compulsory energy efficiency standards governing cooling devices such as ACs, if sufficiently ambitious, will do much to move beyond "business as usual" and help avoid the global climate crisis (Sun and Ferris 2018).

The Dual Strategy to Mitigate Climate Impacts from the Surge in Cooling Demand—Simultaneous HFC Phasedown and Energy Efficiency Improvement

The Montreal Protocol controls about 100 ozone-depleting substances (ODSs) including chlorofluorocarbons (CFCs) and hydrochlorofluorocarbons (HCFCs) that are powerful GHGs. Ozone-depleting HCFCs and ozone-safe HFCs are currently the primary fluorinated refrigerants used in ACs and other refrigeration systems. HFCs were commercialized to rapidly replace ODSs, but are no longer needed because lower global warming potential (GWP)[4] refrigerants delivering higher energy efficiency are available (IPCC 2014; Dreyfus et al. 2020). Most HFCs are super greenhouse gases that can be up to 4,000 times more potent at trapping heat than CO_2.

In October 2016, the Parties to the Montreal Protocol adopted the Kigali Amendment, which expanded the scope of the Protocol to explicitly encompass the phase down of super GHGs, HFCs, although they have only a negligible impact on the ozone layer. Called the Kigali Amendment because it was agreed in the capital of Rwanda, this amendment will avoid nearly 100 billion tons of CO_2-eq by 2050 and up to 0.5°C of warming by 2100. The Kigali Amendment entered into force on January 1, 2019, and a total of 119 countries have ratified the amendment as of April 27, 2021 (Kigali Amendment, UN Treaty Collection website). In addition, President Xi Jinping announced that China will accept the Kigali Amendment in a virtual summit with President Macron of France and Chancellor Merkel of Germany on April 16, 2021 (IGSD 2021).

As background, it is important to recall the Paris Agreement's goal of limiting global warming to "well below 2 °C above pre-industrial levels," aiming for 1.5 °C, and for net zero emissions in the second half of the century (UNFCCC 2015). Researchers calculated that the 1.5 °C threshold could occur up to a decade earlier than predicted in the IPCC Special Report on Global Warming of 1.5 °C, with a 10% chance of reaching it by 2025 (Xu et al. 2018). However, the mitigation potential from the submitted Nationally Determined Contributions will need to be tripled to meet the well below 2 °C goal and increased at least fivefold to meet the 1.5 °C goal (UNEP EGR 2019). Therefore, actions that can magnify climate benefits, like those described below, take on a particular urgency in light of research showing warming was already 1.1 °C above the pre-industrial period in 2016 (WMO 2017).

As the world prepares to redesign refrigerant-using appliances and equipment to replace current HCFC and high GWP HFC refrigerants with more climate-friendly alternatives, there is an opportunity to increase simultaneously the energy efficiency of AC, refrigeration and heat-pump equipment, cutting air pollution and saving consumers money on their electricity bills. According to the quadrennial Scientific Assessment of Ozone Depletion, prepared by the Montreal Protocol's Scientific Assessment Panel, improving the energy efficiency of ACs and other cooling equipment has the potential to double the climate benefits of the Kigali Amendment, meaning that a combined strategy of HFC phasedown and energy efficiency improvement has the potential to reduce projected warming by up to 1 °C by the end of the century (WMO 2018).

China's leadership position in the AC industry calls for action to fulfill the corresponding responsibility to mitigate cooling sector's impact on energy use and climate at both domestic and export markets. For the China market alone, transition of the room AC industry to production of super-efficient and low-GWP ACs by 2030 could provide GHG savings of 850 million tonnes CO_2-eq annually, equivalent to over 8 Three Gorges dams (Shah et al. 2015). A recent analysis finds that targeting continuous improvement through 2030 for room ACs would avoid 49 $GtCO_2$ from energy-related emissions and save consumers USD $6 trillion over 2020–2050 (Phadke et al. 2020). In addition, promoting market penetration of energy-efficient and climate-friendly ACs in both domestic and export markets is in line with China's recent policy and political developments. These include the amendment to the room AC energy efficiency standards (The Minimum Allowable Values of the Energy Efficiency and Energy Efficiency Grades for Variable Speed Room ACs) and the recently announced Belt and Road Green Cooling Initiative (launched during the second Belt and Road Forum for International Cooperation on April 25, 2019) (Ministry of Foreign Affairs of the People's Republic of China 2019). They also include President Xi Jinping's instructions for advancing China's high-quality development and the construction of a global community with a shared future for mankind, and the commitments to raising cooling efficiency in the March and November 2019 bilateral statements between China and France (Ministry for Europe and Foreign Affairs 2019; Xinhua News Agency 2019).

China's Regulatory Scheme for AC Energy Efficiency and the Importance to China of Compulsory Energy Efficiency Standards

China's regulatory scheme for AC energy efficiency includes both compulsory requirements on minimum energy efficiency performance of ACs and incentive mechanisms for promotion of energy-efficient ACs, such as government procurement of energy efficiency products. ACs manufactured in, sold and imported for China must meet the national compulsory energy efficiency standards. Article 25 of the Law on Standardization essentially indicates that "products and services that do not meet compulsory standards shall not be produced, sold, imported or supplied" (Law on Standardization of the People's Republic of China).

China's national government promotes energy-efficient ACs through incentive mechanisms, including the "Energy Efficiency Top Runner" program and government procurement of energy efficiency products. For example, Article 36 of the Environmental Protection Law provides that "[g]overnment agencies and other organizations using fiscal funds shall, when purchasing and using products, equipment and facilities, give priority to those that are energy-saving, water-saving, material-saving or otherwise conducive to protecting environment" (Environmental Protection Law). "Energy Efficiency Top Runner" refers to the best energy-efficient products, companies, or entities (NDRC 2014).

The compulsory energy efficiency standard system has also been undergoing periodical review and amendment to reflect the latest science and technology developments and further transition China's market toward higher efficiency products.[5] In March 2015, the General Office of the State Council issued a document proposing to improve system for review of energy conservation standards and set a target of making more than 80% energy efficiency standard requirements reaching the international leading levels (General Office of the State Council 2015). In November 2015, the National Development and Reform Commission, the Ministry of Industry and Information Technology, and the General Administration of Quality Supervision, Inspection and Quarantine, issued a detailed regulation on the implementation of the room AC Energy Efficiency Top Runner Program, which further clarified that the

Energy Efficiency Top Runner standards will be incorporated into the national compulsory standards based on technology development and market readiness (NDRC 2015).

Strategies for Maximizing the Contribution of Non-industry Nonstate Actors in the Standard Drafting Process

Nonstate actors pursue environmental protection efforts in China through various means described in this book. In terms of national standard making, the roles of nonstate actors in general are enshrined in the Law on Standardization (Law on Standardization of the People's Republic of China). Indeed, this statute specifically promotes the participation of enterprises, social groups, educational and research organizations in standardization work (Law on Standardization of the People's Republic of China). The participation of nonstate actors is strengthened when considered against the backdrop of China's commitments to greater public participation under its World Trade Organization (WTO) accession agreement, including efforts to increase uniform implementation and access to technical regulations and standards (WTO Documents website).

In this section, the authors discuss a series of strategies for maximizing the contribution of nonstate actors in the standard drafting process from the perspectives of an international research institution with the mission of combating the climate emergencies we are facing. Illustrative examples are also demonstrated from the development of China's compulsory room AC standards. Key strategies are described further below and include but are not limited to early involvement in the standard drafting/amendment process (before or after the standard is released for public comment), dynamic strategies to encourage industry stakeholder ambitions based on different companies' development status and preferences, strengthening linkages to market incentive mechanisms including government procurement and fundamental technical and economic assessment support.

Get Engaged as Early as Possible in the Standard Drafting Process

Nonstate actor stakeholder involvement in the national standard-making process differs in terms of the types of involvement and actual impact on the standard draft, depending on whether the nonstate actor is involved *before* or *after* the standard draft is released for public comment. Article 15 of the Law on Standardization provides that during the drafting processes for compulsory and recommended standards, various methods shall be used to solicit comments based on the principles of convenience and effectiveness (Law on Standardization of the People's Republic of China). For the AC standard example under discussion here, which is a national, compulsory standard, the Compulsory National Standard Administration Measures provides that the standard drafting agency shall seek written comments from relevant administrative departments, as well as enterprises, entities, social organizations, consumer organizations, and education and research institutions; post the standard draft on the Internet for public comments (usually for 60 days, but can be shortened to no less than 30 days in case of emergency); and organize meetings and hearings to solicit comments on the compulsory national standards that involve a wide range of concerns and attract broad attention (Compulsory National Standard Administration Measures).

In general, involvement in the standard-making process before the draft is released for public comment guarantees more channels of engagement, including attending standard development workshops and making presentations on proposed standard content, including data supporting increased ambition in the proposed standards, such as raising the level of minimum energy efficiency levels. Involvement prior to release of the draft standard for public comment also ensures that participants gain access to and have the opportunities to comment on the most up-to-date version of standard draft, which is not open to public yet, increasing the potential for proposed adjustments and edits.

After the standard draft is released for public comment, the most significant channel of involvement would be to submit written comments on the draft. The public comment notification is usually a one-time procedure at a stage that typically takes place after most of the standard content has been finalized through consultation with invited experts and industry members. Following the public comment process, the draft standard will be adjusted based on the public comments received, and then submitted for technical expert review and, as needed, notification to WTO Technical Barriers to Trade Committee for WTO member comment, and eventual final approval and promulgation[6] (WTO website; Ferris and Zhang 2002, 2005).

Recognizing that the nonstate actors will have limited opportunities to be involved in the standard-making process after the end of public comment period for draft compulsory standards, getting involved in the process as early as possible as invited experts is a key element of a strategy for maximizing impact. However, a key precondition for early involvement of nonstate actors, particularly those who are not part of the "industry" stakeholders for the standard under discussion, is to get invited as experts to insert themselves in the process before public disclosure of compulsory standards. The interested experts/research institutions need to at least demonstrate that they have established their professional reputation within and outside of China, illustrated their policy or technology advising capacities, and built the necessary personal connections through their research experiences in the relevant fields and/or international project collaboration.

Take the example of the draft amendment to the room AC energy efficiency standard entitled "The minimum allowable values of the energy efficiency and energy efficiency grades for variable speed room ACs" (The Minimum Allowable Values of the Energy Efficiency and Energy Efficiency Grades for Variable Speed Room ACs). The drafting and advising team for the standard was led by the China National Institute of Standardization (CNIS) (a government-affiliated research institution/shi ye danwei) and composed of key, invited industry stakeholders, including Chinese leading AC companies such as Gree, Haier, and Midea, AC component companies such as Sanhua, industry associations such as the International Copper Association (China), and academically oriented nonstate actors, such as the Beijing Industry University and Shanghai Jiaotong University. Nonstate actors invited to join this drafting and advising team enjoy the benefit of early engagement with the draft and invited stakeholder inputs on the draft. Other entities may be able to engage in the process at an early point as invited, contracted experts to provide research and analysis in support of the standard-making/amendment project. This will be discussed in further detail below. In contrast, generally speaking, stakeholders who are not invited to join at this stage will only be able to engage indirectly through their relationships with invited stakeholders, or directly only after the draft standard is released for public comment toward the end of the standard drafting process.

Know and Understand the Industry Stakeholders Engaged in the Standard Drafting Process

The AC industry, including both AC and AC component manufacturers,[7] plays a significant role in the national standard-making process, as one of the key stakeholders in the development of AC energy efficiency standards. In fact, CNIS, the government-affiliated institution in charge of coordinating and administering the standard amendment drafting process, releases the standard amendment draft for public comment after extensive consultation with invited industry stakeholders and experts.[8] This consultation process ensures, in particular from the government agency perspective, that the standard amendment implementation process goes relatively smoothly without major conflict with the industry stakeholders. However, the industry consultation process also presents special challenges to climate mitigation and energy efficiency improvement ambition, since the majority of the industry may in a certain degree favor business-as-usual incremental improvement and hinders the aggressive improvement of energy efficiency requirements that are necessary to meet the climate mitigation and energy saving targets.

This is especially the case in light of the different tiers of AC companies in the industry in terms of technical capacities and company scales. For example, some of these companies[9] endorse the disruptive innovative technology strategies that may significantly reduce climate impacts of the technologies, and others focused on lower-cost technologies and markets with higher climate impacts. Therefore, maximizing success in a non-industry and nonstate actor's strategies to increase climate and energy efficiency ambition of the standards requires more than simply early involvement in the standard drafting process. It also requires a strategy to understand each industry stakeholder's market segments covered in the proposed standards, and what these stakeholder's interests and concerns may be with the standard development. This necessitates interaction with industry stakeholders throughout the standard drafting process, including review of their presentations and consideration of their questions, as well as analysis of publicly available information on the relevant industry sectors, such as through ChinaIOL.com.[10]

In order to best gauge how to increase climate and energy efficiency ambition among the more technically advanced industry stakeholders in the standard drafting process, one should recognize that company market development strategies vary and may influence company preferences with energy efficiency standard requirements. In one example, Company A has a bigger market share in smaller cities and/or less economically-developed regions in China. In this example, strategies aimed at lowering the consumer price of energy efficiency improvements, such as bulk procurement, may need to be prioritized for discussions with Company A assuming Company A has the technical capacity to produce higher-efficiency products. If it turns out that Company A does not have the technical capacity, capacity-building strategies would need to be included.

Non-industry and nonstate actors that wish to raise compulsory standard ambition in order to enhance climate change mitigation could also encourage Company A to support ambition during the standard amendment process through means such as international recognition of their technical advantages. Such efforts should also be complemented with market promotion mechanisms of high energy efficiency products, which would further provide incentives for energy efficiency technology research, development, and deployment, as further discussed below.

In another example, Company B represents a technologically-advanced company that is focused on foreign market expansion. Therefore, Company B responds to arguments to raise ambition to increase access to foreign markets. In this case, standard developments in other countries and by international standard organizations could be of great interest for Company B and can serve as positive motivation for Company B to advance energy efficiency and support higher domestic energy efficiency standards.

In a further example, Company C represents a company that is concerned with access to energy efficiency technologies produced by a supply chain located abroad. In this case, a detailed analysis of supply chain impact and Chinese government policy support on technology innovation would be useful in engaging such a company.

In an additional example, Company D represents a foreign AC company that has an active branch or joint arrangement in China. Company D could play a role during the standard amendment drafting process, after the draft is released for public comments. In particular, Company D could underscore how more stringent standards benefit the company from innovation, competition abroad, and other viewpoints. Such inputs may be most useful in influencing Chinese multinational companies sensitive to competitive forces and whether they are perceived as supporting China innovation and market penetration policies. Depending on where Company D is headquartered and how Company D is organized and managed internally in China, Company D may have different roles to play and various levels of impact on its China branches and/or partners.

Another strategy, aimed at encouraging increased state actor, as well as nonstate industry actor, ambition to accept more stringent and hence, climate-change mitigating, standards, is to have a China law and policy "toolbox" at the ready. To be most effective, "tools" in the toolbox would draw from past law and standard drafting practical experiences. This includes developing a set of Chinese law and policy tools to make such stakeholder acceptance of more ambitious standards easier to happen. For instance, options should be in the toolbox that allow negotiation and implementation flexibility, so that nonstate industry actors will have some extra time to prepare for implementation of higher requirements at the national scale. For example, China's Environmental Air Quality Standard, promulgated in 2012, was implemented in different geographical regions at different times (China Ministry of Ecology and Environment). For instance, in 2012: in Jing-jin-ji (the common reference for the Beijing/Tianjin/Hebei region), the Yangtze River Delta, the Pearl River Delta, Municipalities Directly Under the Central Government (Shanghai, Beijing, Chonqing, Tianjin), and Provincial Capital Cities. In 2013: in 113 "environmental protection key cities" and in "national environmental protection model cities." In 2015: in all cities above the prefectural level. And on Jan. 1, 2016: nationwide.

Engage Early and Credible Research Institution Support in the Standard Drafting Process

Early and credible support of nonstate actor standard-drafting recommendations from both foreign and domestic technical research institution experts is also critical in order for non-industry and nonstate actors to maximize their contribution to raising the climate and energy efficiency ambition of national standards. This is particularly helpful for standard drafting proposals involving technical issues. The selection of such technical institutions/experts is key to ensuring that the industry experts are respected and experienced experts in areas relevant to the standard in question. Hence, the organization overseeing the standard drafting process, for instance CNIS, should also be comfortable with the qualifications and participation of the technical research institution(s) in question.

The experts from the technical research institutions are typically tasked with evaluating the technical concerns and issues raised during the standard-making discussions, and provide authoritative conclusions on key issues. These issues include:

1. whether technical questions and concerns are actually technical challenges or economic concerns disguised as technical challenges; and
2. what the potential pathways are to solve the actual technical challenges and move toward the direction of technically feasible and environmentally friendly standards, which are in line with Chinese national policy priorities of Constructing an Ecological Civilization, the Made in China 2025 initiative and the Construction of a Green Belt and Road.

In practice, foreign/international research institutions and laboratories can play a particularly important technical supporting role when they collaborate with China-based research institutions and laboratories. For example, foreign/international research institutions can provide valuable information and inputs to the Chinese government and affiliated entities on ambition-raising foreign technologies, market prevalence of advanced technologies, and greater stringency (higher minimum energy efficiency thresholds) as important references guiding discussion away from mere "incremental" change in standards, during standard drafting discussions. In light of China's emphasis on brand image improvement abroad and technology innovation, this can be a particularly compelling means for the standard-drafting authorities to show government peers that they are acting consistent with government policy aims, as well as to the industry stakeholders who wish to compete outside of China on the global market.

Link Ambition-Raising Proposals to Market Incentive Mechanisms

Market incentive mechanisms play a significant role in promoting energy efficiency technology research, development and deployment, and consequently, are considered key components of the energy efficiency standard implementation system. As we described above, the energy efficiency standards set up the energy-efficient product levels (usually referring to the grade 1 and grade 2 of the energy efficiency requirements), as well as the minimum energy efficiency grade (lowest energy efficiency grade provided in the standard). In addition, grade 1 and grade 2 requirements of the standards are also linked to government green procurement requirements and product catalogs, which have significant impact on the market penetration of energy-efficient products (General Office of the State Council). The national and local government subsidy programs mentioned above usually apply on grade 1 and grade 2 products as well (Beijing Municipal Bureau of Commerce).

Market buying power, if used properly, can effectively guide and incentivize industry technology innovation and promote market transition and society development (Nikit et al. 2017). The highest energy efficiency level (i.e., a "top runner") would in particular send the signal for innovation toward highest efficiency, and chart the path for China's leadership in producing the technologies of the future. Therefore, non-industry and nonstate actors wishing to maximize raising the climate and energy efficiency ambition of national AC energy efficiency standards should focus on phasing out inefficient products through the provision of the minimum energy efficiency level. They should also consider advocating that a grade-1 (top) level be decided based on the best energy performance product available on the market. In this way, it would be a technically feasible while also ambition-raising exercise for AC manufacturers to meet the requirements.

In addition, such stakeholders should make note of the history of AC standard development and strengthening and how this has shown that the AC prices always go down as the standard becomes more stringent and market share increases (Nikit et al. 2017). Enterprises taking the opportunity to lead on energy efficiency improvement would secure their market competitiveness in both domestic and export markets, and strengthen their company strategic development in the longer term.

Conclusions

This chapter lays out the critical role of cooling efficiency in solving the climate crisis we are facing while ensuring human wellness and social prosperity. In particular, China as the world's largest manufacturer, consumer, and exporter of ACs is in a unique leadership position on this issue. As the world is getting started or getting ready to phase down the climate damaging HFCs, an opportunity also presents for countries, including China, to simultaneously improve AC energy efficiency during refrigerant transition. In addition to the governments, nonstate actors and stakeholders can and have been involved in these important processes. For China, compulsory energy efficiency standards are critical parts of the country's policy and regulatory scheme on AC and other electronic product energy efficiency performance. This, and our particular experience with the energy efficiency standard-making process, are reasons why we draw from the recent AC energy efficiency standard revisions to illustrate key strategies to maximize non-industry and nonstate actor contributions to the national standard drafting and amendment.

In particular, we recommend, for nonstate actors advocating for raising the energy efficiency and climate mitigation ambition of such national standards, the following key strategies:

- Get involved early in the standard drafting process (meeting the preconditions to get invited as credential technical/policy experts);
- Study and understand the industry related to the drafted standard (including but not limited to the technical capacities, market assessments, and company development strategies of different industry stakeholders involved in the standard drafting process);
- Obtain credible and early research institution support in the standard drafting process (both domestic and foreign technical research institutions have important roles to play in this process, noting the particular strength of foreign technical research institutions that collaborate well with local Chinese research institutions and laboratories); and
- Link your ambition-raising proposals to market incentive mechanisms.

Appendix I: Different Types of Standards in China as Provided in the Law on Standardization

Article 2 of the Law on Standardization of the People's Republic of China is the starting place for understanding China's complex standardization system, including officially recognized standard nomenclature and the legal force of various types of standards. This law provides that "[s]tandards include national standards, industry standards, local standards, group standards, and enterprise standards." As also provided in Article 2 of the Law on Standardization, national standards can be either compulsory or recommendatory. Industry standards and local standards are all recommendatory. As for group standards and enterprise standards, while the National Standardization Law recognizes these categories of standards, it is noteworthy that they are not within the category of national standards that have officially sanctioned compulsory or recommendatory norm-setting status. Instead, these standards are for adoption within certain groups or enterprises. An exception is that group standards can be made open for voluntary adoption outside of the group according to the group's rules.

Further clarification and explanation on the meaning of "groups" and "group standards" can be found in Article 18 of the Law on Standardization, which provides that "[t]he State encourages social organizations such as societies, associations, chambers of commerce, federations, industrial technology alliances, etc., to coordinate with relevant market entities to jointly formulate group standards that meet the needs of the market and innovation. [These group standards] can be agreed upon and adopted by members of the group or used voluntarily by society in accordance with the rules of the group."

Another useful provision to keep in mind is Article 21 of the Law on Standardization, which sets compulsory national standards as the "baseline" that other referenced standards must maintain or ideally exceed. Article 21 provides that "[t]he technical requirements of recommended national standards, industry standards, local standards, group standards, and enterprise standards must not be lower than those of compulsory national standards. The state encourages social groups and enterprises to formulate group standards and enterprise standards that have higher technical requirements than the recommended standards."

Appendix II: Implementation Watch-List for Nonstate Actors Following Compulsory National Standard Promulgation

1. Full text of the new standard is posted on the Web site: National Standard Full Text Disclosure System, within 20 days after promulgation of the standard (Article 37 of the Compulsory National Standard Administration Measures);
2. The new standard is enforceable from the effective date identified in the standard (older version of the standard expires simultaneously) (Article 39 of the Compulsory National Standard Administration Measures) if no standard implementation delay is formally announced; (e.g., the Standardization Administration of China announced on April 1, 2020 a batch of 13 standards subject to a six-month implementation delay due to the impact of COVID-19 epidemic);
3. For compulsory energy efficiency standards, manufacturers and importers use the (applicable) updated China Energy Label (Energy Efficiency Label Management Regulation [2016]);
4. Interpretations to the new standard are issued according to law, which have equal legal validity as the new standards (Article 41 of the Compulsory National Standard Administration Measures);
5. Feedback on new standard implementation is submitted through the National Standard Full Text Disclosure System (Article 42 of the Compulsory National Standard Administration Measures).
6. The standard drafting agency continually monitors and evaluates implementation status of the new standard and conducts a standard review in no more than five years from when the new standard was promulgated (Articles 43–45 of the Compulsory National Standard Administration Measures). Note: energy efficiency standards have a shorter review cycle indicating a standard review in **no more than three years** from when the new standard was promulgated (General Office of State Council, Opinions on Strengthening the Standardization of Energy Saving [2015] [emphasis added]).

Notes

1. This chapter, as indicated here, only discusses the nonstate actor's participation in the national standard drafting process. Of course, the standard drafting processes naturally includes some considerations related to promulgation and implementation of the standard under discussion. However, this chapter does not cover the certification and accreditation, which are separate and complicated topics by themselves.
2. *See* Appendix I on the different types of standards in China as provided in the Law on Standardization.
3. The export of ACs also plays a significant role within the global sales of ACs. For further introduction on the issue of environmental dumping, see Dr. Stephen O. Andersen et al. 2018. Defining the Legal and Policy Framework to Stop the Dumping of Environmentally Harmful Products. *Duke Environmental Law & Policy Forum* 29: 1–48. https://scholarship.law.duke.edu/delpf/vol29/iss1/1.
4. GWP is "[a]n index measuring the radiative forcing following an emission of a unit mass of a given substance, accumulated over a chosen time horizon, relative to that of the reference substance, carbon dioxide (CO_2). The GWP thus represents the combined effect of the differing times these substances remain in the atmosphere and their effectiveness in causing radiative forcing." See Intergovernmental Panel on Climate Change (2014).
5. See Appendix II for an implementation watch-list for nonstate actors following compulsory national standard promulgation, including periodical review.
6. This is an obligation of all WTO members, China included. China's obligations were established as part of China's accession to the WTO and related Accession Agreement (2001). For background, see Ferris, R., and Zhang, H. (2002, 2005).
7. Note that some of the enterprises are state-owned enterprises, so they are not considered nonstate stakeholders in that way.
8. Although keeping in mind that high-level political instructions and national government agency involvement also have strong influence on how this consultation process goes. *See* a list of AC industry stakeholders who were directly involved in the AC energy efficiency standard amendment process as key members of the standard drafting team: https://members.wto.org/crnattachments/2019/TBT/CHN/19_5317_00_x.pdf.

9. Gree Electric Appliances Inc. was selected as one of the eight finalists for the Global Cooling Prize, an international innovation competition to develop super-efficient and climate-friendly residential cooling solutions for homes. World's leading AC manufacturers and innovative technology companies in running for the Global Cooling Prize. 2019. See https://globalcoolingprize.org/finalist-global-press-release/.
10. ChinaIOL.com is an online platform that provides China manufacturing industry information and data collection and analysis.

References

Beijing Municipal Bureau of Commerce, Beijing Municipal Bureau of Finance, Beijing Development and Reform Commission, and Beijing Bureau of Water Affairs. 2019. Notice on Implementation of the Energy Saving, Emission Reduction and Consumption Promotion Policies. Beijing. http://sw.beijing.gov.cn/zwxx/zcfg/gfxwj/201912/t20191221_1388239.html.

China Ministry of Ecology and Environment. 2012. Circular on the Implementation of the Environmental Air Quality Standard (GB3095-2012). http://www.gov.cn/zwgk/2012-03/02/content_2081004.htm.

China's Draft Room AC Standard Amendment for WTO Notification. 2020. https://members.wto.org/crnattachments/2019/TBT/CHN/19_5317_00_x.pdf.

Compulsory National Standard Administration Measures. Promulgated on 6 January 2020 and effective as of 1 June 2020. http://gkml.samr.gov.cn/nsjg/fgs/202001/t20200113_310467.html.

Dreyfus, Gabrielle B., Nathan Borgford-Parnell, John Christensen, David W. Fahey, Brian Motherway, Toby Peters, Romina Picolotti, and et al. 2020. Assessment of Climate and Development Benefits of Efficient and Climate-friendly Cooling. Paris: CCAC. https://ccacoalition.org/en/resources/assessment-climate-and-development-benefits-efficient-and-climate-friendly-cooling.

Environmental Protection Law. Promulgated on 24 April 2014 and effective as of 1 January 2015. http://www.gov.cn/zhengce/2014-04/25/content_2666434.htm.

Ferris, Richard, and Hongjun Zhang. 2002. The Challenges of Reforming a Legal Culture: Assessing the Status Quo and Looking at Post-WTO Admission Challenges for the People's Republic of China. *The Georgetown International Law Review* 14 (3): 429–460.

Ferris, Richard, and Hongjun Zhang. 2005. Environmental Law in the People's Republic of China: An Overview Describing Challenges and Providing Insights for Good Governance. In *China's Environment and the Challenges of Sustainable Development*, ed. Kristen A. Day. Armonk, NY: M.E. Sharpe.

General Office of the State Council. 2007. Circular on Establishing Government Compulsory Energy Saving Product Procurement Mechanism. http://www.gov.cn/zwgk/2007-08/06/content_707549.htm.

General Office of the State Council. 2015. Opinions on Strengthening the Standardization of Energy Conservation. http://www.gov.cn/zhengce/content/2015-04/04/content_9575.htm.

Intergovernmental Panel on Climate Change. 2014. Climate Change 2014: Synthesis Report. Contribution of Working Groups I, II and III to the Fifth Assessment Report of the Intergovernmental Panel on Climate Change, 117–130 [Core Writing Team, R.K. Pachauri and L.A. Meyer (eds.)]. Annex II: Glossary [Mach, K.J., S. Planton, and C. von Stechow (eds.)]. Geneva: IPCC.

International Energy Agency. 2018. The Future of Cooling. Paris: IEA. https://www.iea.org/cooling/.

International Energy Agency. 2019. The Future of Cooling in China. Paris: IEA. https://www.iea.org/reports/the-future-of-cooling-in-china.

Law on Standardization of the People's Republic of China. Promulgated on 4 November 2017 and effective as of 1 January 2018. Articles 15 and 22. http://www.npc.gov.cn/zgrdw/npc/xinwen/2017-11/04/content_2031446.htm.

Minimum Allowable Values of the Energy Efficiency and Energy Efficiency Grades for Variable Speed Room Air Conditioners. Promulgated on 31 December 2019 and effective as of 1 July 2020. http://openstd.samr.gov.cn/bzgk/gb/newGbInfo?hcno=BC04CDC71AD8C36B62C0FF4AE58F633C.

Ministry for Europe and Foreign Affairs. 2019. China-France Beijing Call for Biodiversity Conservation and Climate Change. https://www.diplomatie.gouv.fr/en/french-foreign-policy/climate-and-environment/news/article/beijing-call-for-biodiversity-conservation-and-climate-change-06-nov-19.

Ministry of Foreign Affairs of the People's Republic of China. 2019. List of Deliverables of the Second Belt and Road Forum for International Cooperation. https://www.fmprc.gov.cn/mfa_eng/zxxx_662805/t16

58767.shtml (*note*: Initiative 9 of the Initiatives Proposed or Launched by the Chinese Side).

National Development and Reform Commission, Ministry of Finance, Ministry of Industry and Information Technology, Government Offices Administration of the State Council, National Energy Administration, General Administration of Quality Supervision, Inspection and Quarantine, and National Standardization Administration. 2014. Implementation Plan for Energy Efficiency Top Runner Program. http://www.miit.gov.cn/n1146285/n1146352/n3054355/n3057542/n3057544/c3634683/content.html.

National Development and Reform Commission, Ministry of Industry and Information Technology, and General Administration of Quality Supervision, Inspection and Quarantine. 2015. Implementation Regulation on Energy Efficiency Top Runner Program for Variable Speed Room Air Conditioner. https://www.ndrc.gov.cn/xxgk/zcfb/tz/201511/t20151104_963492.html.

Nikit, Abjyankar, Nihar Shah, Won Young Park, and Amol A Phadke. 2017. Accelerating Energy Efficiency Improvements in Room Air Conditioners in India: Potential, Costs-Benefits, and Policies. Berkeley: LBNL. https://ies.lbl.gov/publications/accelerating-energy-efficiency.

Phadke, Amol, Nihar Shah, Jiang Lin, Won Young Park, Yongsheng Zhang, Durwood Zaelke, Chao Ding, Nihan Karali. 2020. Chinese Policy Leadership Would Cool Global Air Conditioning Impacts: Looking East. *Energy Research & Social Science* 66. https://www.sciencedirect.com/science/article/pii/S2214629620301468.

Protocol on the Accession of the People's Republic of China. Decision of 10 November 2001 (WT/L/432), Article 2(C) (Transparency). https://www.wto.org/english/thewto_e/acc_e/completeacc_e.htm.

Shah, Nihar, Max Wei, Virginie Letschert, and Amol Phadke. 2015. *Benefits of Leapfrogging to Super efficiency and Low Global Warming Potential Refrigerants in Room Air Conditioning*. Berkeley: Lawrence Berkeley National Laboratory. https://eta.lbl.gov/sites/default/files/publications/lbnl-1003671.pdf.

Sun, Xiaopu, and Tad Ferris. 2018. The Kigali Amendment's and China's Critical Roles in Evolving the Montreal Protocol. American Bar Association (ABA). http://www.igsd.org/the-kigali-amendments-and-chinas-critical-roles-in-evolving-the-montreal-protocol/.

Sustainable Energy for All and Kigali Cooling Efficiency Program. 2019. *Chilling Prospects: Providing Sustainable Cooling for All*. Vienna: SEforALL,

KCEP. https://www.seforall.org/sites/default/files/SEforALL_CoolingForAll-Report.pdf.

United Nations Environment Programme. 2019. Emissions Gap Report 2019. Nairobi: UNEP. https://www.unep.org/resources/emissions-gap-report-2019.

United Nations Environment Programme and International Energy Agency. 2019. *Cooling in A Warming World – Opportunities for Delivering Efficient and Climate Friendly Cooling for All*. Nairobi: UNEP, IEA. http://coolcoalition.org/cooling-in-a-warming-world-opportunities-for-delivering-efficient-and-climate-friendly-cooling-for-all/.

United Nations Framework Convention on Climate Change. Conclusion date: 12 December 2015. Adoption of the Paris Agreement, FCCC/CP/2015/L.9 (2015).

United Nations Treaty Collection. Kigali Amendment Open Climate Data. Last updated 27 April 2021. https://kigali-amendment.openclimatedata.net/.

World Meteorological Organization. 2017. *WMO Statement on the State of the Global Climate in 2016*, 4. Geneva: WMO.

World Meteorological Organization. 2018. "Scientific Assessment of Ozone Depletion: 2018. Global Ozone Research and Monitoring Project – Report No. 58," 588. Geneva: WMO.

Xinhua News Agency. 2019. "China-France Joint Declaration on Maintaining Multilateralism and Improving Global Governance." Mar 26, 2019. http://www.gov.cn/xinwen/2019-03/26/content_5377035.htm?from=groupmessage&isappinstalled=0.

Xu, Yangyang, Veerabhadran Ramanathan, and David G. Victor. 2018. Global Warming Will Happen Faster Than We Think. *Nature* 564: 30–32. https://www.nature.com/articles/d41586-018-07586-5.

In the Shadow of the State: The Rise and Limits of Transnational Private Certification in China's Seafood Sector

Yixian Sun

Introduction

Today, non-state actors—including businesses and civil society groups—play a prominent role in tackling a number of environmental problems ranging from deforestation and fisheries depletion to air pollution and climate change. In many instances, cross-border networks were formed to authoritatively steer their constituents toward public goals—a phenomenon conceptualized as "transnational governance" (Andonova et al. 2009; Hale and Held 2011).[1] When these governance institutions do not derive their authority directly from sovereign states, intergovernmental bodies, or subnational governments, researchers describe them as "private" or "non-state" arrangements (e.g., Cashore 2002; Bartley 2007; Auld 2014).[2] Many such private systems set voluntary standards

Y. Sun (✉)
University of Bath, Bath, UK
e-mail: ys2327@bath.ac.uk

D. Guttman et al. (eds.), *Non-state Actors in China and Global Environmental Governance*, Governing China in the 21st Century,
https://doi.org/10.1007/978-981-33-6594-0_9

and use global supply chains to recognize, track, and label products and services from responsible businesses (Bernstein and Cashore 2007). These systems are able to link buyers and Western consumers to producers in developing and emerging economies by disclosing information about producers' practices. The rise and growth of these governance schemes (especially eco-labels) has been remarkable in the last fifteen years and has sparked an abundant literature. However, China—the largest producer and consumer of many commodities targeted by private certification and labeling schemes—remains largely missing in this literature.[3] Without examining China's involvement in this new governance mode, we cannot determine whether these initiatives are likely to make significant contributions to the sustainability of the Earth system.

Past studies on transnational sustainability governance focus primarily on the rise of relevant schemes in developed countries, namely in Western Europe and North America where most schemes first emerged (e.g., Cashore et al. 2004; Bartley 2007; Gulbrandsen 2009). Much less attention has been given to the dynamics around the diffusion, operation, and impact of transnational private governance in developing and emerging economies, although in most cases these countries experience more negative impacts of the global economy than wealthy countries (van der Ven et al. 2021). Accordingly, there is an urgent need to advance our understanding of how transnational private governance initiatives operate in economic and sociopolitical systems other than advanced democracies.

China is a critical case among developing countries and emerging economies due to its unique institutionalized processes of environmental governance. Dominant environmental governance processes in China include state planning and government responses to crises, which make China distinct from Western democracies like the US (Young et al. 2015). However, it remains unclear whether non-state actors have little role to play in addressing the environmental challenges facing China or if a different governance process led by non-state actors has emerged in the country to bridge governance gaps left by the actions of the state. As suggested by Guttman et al. (2018) and other chapters in this volume, answering this question with regard to China requires us to rethink and re-conceptualize the distinction between the state and non-state actors in China's governance system.

This chapter addresses this issue by investigating how transnational private governance, in the form of certification and labeling schemes, has been reshaped by the landscape of environmental governance in China. Studying these schemes requires us not only to examine international non-state actors who manage or support these schemes but also Chinese non-state and state actors who interact with them. While sustainability standards and certifications have become popular in many agri-food value chains, seafood is a particularly important case for understanding the operation and influence of such private governance schemes and non-state actors in China. In the last decade, non-state actors created several standards, certifications, and labels to promote sustainable seafood production and consumption, and they have rapidly increased their market uptake across the globe (Jacquet et al. 2010; Potts et al. 2016). Many of them have also attempted to expand their activities in China (Fabinyi 2016). Yet, until recently, these private governance initiatives have been marginal in China's seafood sector, though the country is the world's largest producer and consumer of seafood (Potts et al. 2016). To better understand the potential and challenges of private seafood governance in China, we need to identify the actors who have promoted relevant initiatives, examine how they have interacted with influential domestic stakeholders, especially state agencies, and assess the extent to which these initiatives complement public regulations to improve the sustainability of China's seafood industry. Answering these questions requires us to consider the complex interaction between international and Chinese actors as these private initiatives can be shaped by and also reshape domestic governance in China.

To investigate the operation and impacts of private seafood certifications in China, I draw on data gathered from field interviews, policy documents, and academic literature. The following analysis shows that transnational private initiatives can sow their seeds in China if they strategically align themselves with the state's policy goals and engage with actors that are central to domestic governance processes. Nonetheless, the evidence to date suggests that while private governance has introduced some new concepts and practices to Chinese stakeholders, they are far from moving the whole sector toward sustainability.

A Framework Assessing the Influence of Transnational Private Governance in China

Examining the influence of transnational private governance requires researchers to answer two sets of questions. First, given multiple actors involved in such initiatives, who are the major stakeholders and what are their incentives to support (or not) relevant rules developed by non-state actors? Second, from a problem-solving perspective, to what extent can they contribute to solving or mitigating the problems that motivate those who create them (Young 2011)?

Mapping Key Stakeholders

To understand the operation of private certification schemes in China, we need to consider not only who developed and who supports them, but also with whom their supporters interact to implement their rules in the local context. I draw on the framework proposed by the book's introductory chapter and Guttman et al. (2018) to identify key actors that may influence the functioning of transnational private governance in China (see Fig. 1).[4]

First, companies and environmental NGOs based in developed countries may want to diffuse the rules and standards that they created into China. But past research has suggested that environmental activism in China is "embedded" in the sense that there is "a negotiated symbiosis between Party, state, and society, whereby the Party-state draws certain boundaries" (Ho 2007: 37). As a result, in China transnational NGOs are more likely to rely on non-confrontational strategies to promote private governance.[5]

Second, domestic regulators have the means to enhance or undermine the authority of transnational private governance constitute another group of stakeholders. On the positive side, the state can lend its support to private governance by removing policy barriers and providing training or subsidies to compliant companies (Lister 2011; Gulbrandsen 2014). On the negative side, the state can delegitimize private governance organizations by restricting their activities within its borders or developing

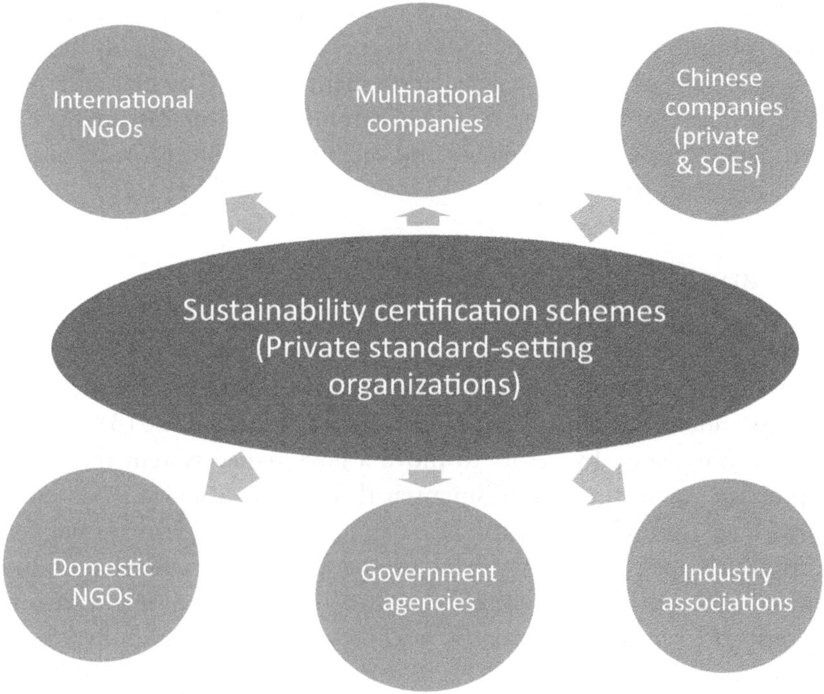

Fig. 1 Actors interacting with private certification schemes in China

competing schemes (Schouten and Bitzer 2015; Wijaya and Glasbergen 2016). In China's political context, state regulators are likely to play a prominent role in conditioning governance arrangements of non-state actors (van Rooij et al. 2016). This means that how certification schemes interact with the Chinese state and shape the latter's interest is critical to how their rules are disseminated.

Additionally, private governance organizations may interact with other domestic actors such as business associations and research institutes (often called *shiye danwei* or *shiye tuanti* meaning "public service units").[6] While analogous organizations are usually considered independent from the state in Western democracies, their Chinese counterparts function differently due to their links to the government (see the book's introductory chapter). Yet, these actors may still develop their own organizational interests and be able to motivate policy change by influencing policymakers (Mertha 2009; Teets 2014, 2018). The influence of these associations may be particularly salient on state-owned enterprises (SOEs). Therefore, support from such domestic actors can increase the legitimacy of private governance and facilitate the diffusion of relevant knowledge and practices.

Evaluating the Performance

Research on regime effectiveness suggests dealing with the question of governance performance by assessing three indicators at distinct levels: outputs, outcomes, and impacts (Young 1999, 2011, 2013). Outputs are measures taken by actors to move a governance system from paper to practice. Accordingly, assessment at this level focuses on what various actors have done in China to ensure functioning of transnational certification schemes. Outcomes refer to changes in the behavior of those subject to a regime's provisions that can be attributed to the regime's operation. To measure outcomes, we need to consider carefully whether the actors' behavioral changes were due to newly-adopted private rules or standards. Lastly, impacts concern the extent to which a governance system plays a role in mitigating the problem it intends to address. Measuring impacts is a challenging task as many outcome variables are predetermined by other forces (e.g., geography, level of development, and cultural norms) that may interact in complex ways with a governance intervention. Additionally, most governance schemes need some time to have discernable impacts on ecosystems such that researchers have difficulty getting accurate data to measure variables of interest. For these reasons, I leave assessment at this level for future research.

Methodology

I employ a qualitative approach to empirically assess the role of transnational private governance in China's seafood sector. Many seafood certification schemes have proactively promoted their standards in China, and, accordingly, constitute a suitable testing ground for this new mode of governance in the Chinese context. Moreover, due to verification requirements and labels being directly shown to consumers, these schemes interact with multiple Chinese stakeholders. Analyzing such interaction can therefore advance our understanding on the promise and limits of transnational private governance in China. By tracing the historical processes through which relevant schemes entered China and expanded their activities, this analysis has two goals. First, it aims to identify the diffusion mechanisms of transnational private governance in China to understand dynamic interaction between state and non-state actors. Second, it assesses, in a preliminary way, the influence of the relevant schemes in promoting sustainable seafood, and therefore, sheds light on the promise of transnational private governance in China's governance processes. The data are drawn from various sources. In addition to secondary data, such as reports by relevant organizations and academic publications, I also use interview data gathered from fieldwork in 2017 for my doctoral dissertation (Sun 2018).

The Rise of Private Sustainability Certification in China's Seafood Sector

For centuries, seafood has been a critical food source of the global population.[7] The global seafood industry has experienced rapid expansion over the last 50 years due to a continuous increase in fish consumption around the world (FAO 2016). As one of the most-traded food commodities worldwide, the sector also provides livelihoods for millions of fishermen and fish farmers. For both wild-caught and farmed seafood, China plays a predominant role in today's global seafood supply chain as the largest producer and consumer. As of 2015, China represented nearly 20% of global capture fisheries production and over 60% of global aquaculture production. Over the last decade, China has also become a major seafood importer due to a growing domestic market (FAO 2016).

Both the subsectors of wild catch fisheries and aquaculture face significant sustainability challenges, including the decline and collapse of species due to over-exploitation, water and soil pollution caused by the use of antibiotics and drugs, and introduction of invasive species to local waters, as well as social issues such as forced labor (Pikitch 2012).[8] In this context, non-state actors created voluntary governance schemes to promote sustainable seafood production and consumption. In the last two decades, these programs quickly increased their market uptake such that the relevant rules and standards were introduced to producers around the world (Gulbrandsen 2009; Auld 2014; Potts et al. 2016). Given the size of the Chinese seafood industry and market, supporters of sustainable seafood have shown a strong interest in promoting these initiatives in China and have made efforts to engage with the relevant stakeholders.

The following analysis focuses on three private seafood certification schemes: the Marine Stewardship Council (MSC), Aquaculture Stewardship Council (ASC), and Best Aquaculture Practices (BAP). The MSC governs wild capture fisheries whereas the other two schemes set standards for aquaculture. These schemes are transnational multi-stakeholder initiatives that set process standards on sustainable seafood and use third-party auditing to verify compliance. In addition to playing important roles in the global seafood market, all three schemes were introduced to China's seafood industry a decade or so ago and currently operate offices in China. On their legal status, they are usually NGOs or non-profit groups in the West, but in China they may register as corporations to conduct activities due to the government's tight control over foreign NGOs.

Marine Stewardship Council

In the late 1990s, the Marine Stewardship Council (MSC) emerged as the first certification program targeting the seafood sector. It resulted from a partnership between environmental NGOs and multinational companies led by the World Wide Fund for Nature (WWF). This certification scheme set standards for sustainable management for wild capture fisheries primarily to address the issue of overfishing (Gulbrandsen 2009; Auld 2014). To ensure product traceability, the program also issues chain-of-custody (CoC) certificates for businesses in seafood supply chains, such as possessors and retailers. With the support of large retailers in Europe and North America, the MSC has become the most well-established eco-certification program in the global seafood market and, to date, its certified production accounts for nearly 15% of marine wild catch in the world.[9]

In the mid-2000s, MSC certification was introduced to China's seafood industry, particularly targeting processing companies in the middle of the global supply chain. Initially, the program gained the attention of Chinese processors due to the sourcing requirements of large retailers and seafood brands in developed markets.[10] For instance, in Europe, Sainsbury's committed in 2003 to stocking only sustainable wild catch by 2010, and in North America, Walmart joined the movement by announcing a commitment in 2006 to purchase all wild-caught fresh and frozen fish for the US market from MSC-certified fisheries within the next three to five years (Walmart 2006). As a result, Chinese seafood processors exporting to developed countries were soon affected by this new market orientation toward sustainable products and began to take up the MSC CoC certification in order to maintain orders from foreign buyers.[11] Figure 2 shows that since 2006 more and more companies in China have adopted the MSC CoC standard to prove the traceability of

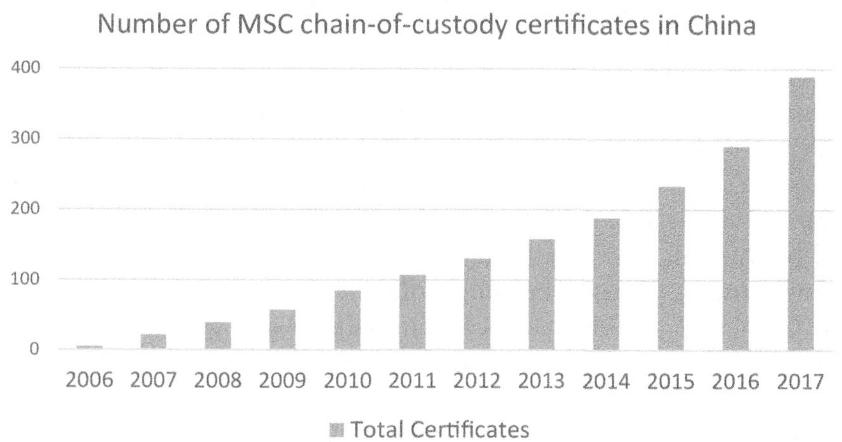

Fig. 2 Growth of the MSC chain-of-custody certification in China (This is a modified version of the graph in Sun [2018: 81], data source: the MSC website)

seafood from certified fisheries. Most of these are processing companies. As of September 2019, the total number of the MSC CoC certificates in China surpassed 450.

Yet, the dissemination of the MSC CoC certificates through global supply chains may not be sufficient to significantly improve the sustainability of Chinese fisheries. To comply with the CoC standard, companies only need to establish a traceability and management system to be able to identify the sources of their supply from sustainably managed fisheries.[12] For most of these processors, only part of the seafood they deal with is from certified fisheries and most of certified products are sold abroad instead of in China's domestic market. In other words, the support for the MSC among Chinese processors focusing on re-exporting business may not produce the rise of a sustainable seafood movement within China.

For quite long time, MSC-certified products remained marginal in the Chinese seafood market: when the program set up an office in Beijing in 2013, there were only three to five types of MSC-labeled products present in the domestic market—all were imported and made by foreign brands.[13] But the MSC has made rapid progress over the last five to six years in gaining the support of Chinese stakeholders and introducing certified products to Chinese consumers. We can observe this trend in Fig. 2 through the increasingly faster growth in the number of CoC certificates after 2013.

As explained below, this momentum can be attributed to the MSC's efforts of engaging influential domestic stakeholders in China's seafood industry and raising their awareness about sustainable seafood. In fact, when the program decided to expand its activities in China in the early 2010s, its leadership paid close attention to China's governance processes led by the state and eagerly sought collaboration with relevant government agencies. From the beginning of their operation in China, they told the Chinese regulators that they "want[ed] to work with them and help them."[14] Sharing a strong willingness to collaborate, the MSC found some central industry associations in China which had been interested in using sustainability certification to reform China's seafood supply chain. As discussed in section "A Framework Assessing the Influence of Transnational Private Governance in China", industry associations in China may be extensions of some state agencies and therefore can have strong influence on the policy-making process by bridging the space between the state and the market. Accordingly, partnerships with relevant associations provide the MSC with the opportunity to expand its influence in China.

In this respect, a key partner of the MSC is China Aquatic Products Processing and Marketing Alliance (CAPPMA), one of the most influential associations in China's seafood industry. Under the direction of the Ministry of Agriculture, CAPPMA, as a *shiye tuanti*, has been the only representative of the seafood industry at the national level to serve as an intermediary between government and businesses. Its members comprise major companies engaging in activities along the seafood value chain, and the state has delegated regulatory authority in several areas to this association, including proposing and designing standards on product quality and supervising the seafood export.[15] Thus, for most companies in China's seafood sector, CAPPMA provides trustworthy information and signals the direction of government policies.

CAPPMA has treated transnational certification schemes like the MSC as useful strategies to promote sustainable fisheries management in China, which is a central goal set by the Chinese state for its fisheries policy (Cao et al. 2017; Cui 2015). To introduce sustainability certification to Chinese stakeholders, for a decade CAPPMA has organized the annual "Sustainable Seafood Forum" in conjunction with the China Fisheries and Seafood Expo. The forum was initially suggested by WWF, a founder of the MSC, which was seeking partners in the Chinese government to raise awareness about sustainable seafood in the country. Realizing CAPPMAs willingness to promote sustainability certification, WWF-China established a partnership with the association, and the MSC and the ASC to organize this annual event. Through CAPPMA, the forum has been able to invite many key stakeholders in China's seafood governance including major domestic seafood producers as well as officials from the Ministry of Agriculture and the Certification and Accreditation Administration (CNCA).[16] Accordingly, it provided a platform for the MSC and its supporters to explain sustainable fisheries management and its benefits to potential adopters in China.

Moreover, the MSC has actively introduced its standards and labels to downstream businesses and consumers in China's domestic market. In this respect, the program again made progress by engaging with a central industry association (also labeled as a *shiye tuanti* in China's political system) in China's retail sector—the China Chain Store and Franchise Association (CCFA). Since 2014, based on this partnership, the MSC has been able to participate in the annual "Sustainable Consumption Week," jointly organized by CCFA and WWF through marketing campaigns in supermarkets and e-commerce platforms (MSC 2017). Through such promotional activities, more and more consumers have become aware of the MSC label and many of them also expressed willingness to pay some price premiums for sustainably produced seafood (Li et al. 2017; Goal-Blue 2019). Additionally, through the CCFA, the MSC has proactively approached major retailers in China, especially online platforms that have risen as key channels of seafood purchase in China. A key milestone is the commitment of Tmall—one of the largest e-commerce platforms in China owned by the Alibaba group—to sell 20% of its seafood with the MSC label by 2020 (MSC 2017). With growing support of retailers, it is realistic to expect continuous increase in the sale of MSC-certified products in China.

To summarize, while the MSC certification was initially introduced to Chinese companies by Western buyers more than a decade ago, only recently and with the support of some industry associations has the program begun to rise in China's domestic market. As these associations are sponsored by the government and play a key role in supervising and coordinating the relevant industries, establishing collaboration with them has been helpful in building the program's legitimacy in China and raising awareness of Chinese businesses about sustainable seafood. Unlike their Western counterparts, Chinese industry associations serve as state actors that represent the government and exercise strong influence on the market. Instead of using boycotts or naming and shaming campaigns, the MSC has proactively sought partnerships with government agencies or industry associations to expand its activities in China. Despite the progress discussed above, we must also recognize that the MSC has made little contribution to change practices of Chinese fisheries as only one fishery in China's territorial waters is MSC-certified. This limitation is discussed in section "Outputs and Outcomes of Transnational Private Governance in China".

Aquaculture Stewardship Council

Following the MSC's success, in 2010 WWF and the Dutch Sustainable Trade Initiative created the Aquaculture Stewardship Council (ASC) as a certification program for farmed seafood. Adopting a model similar to that of the MSC, the ASC awarded its first certificate to a tilapia farm in 2012. Since then, the program has increased its market uptake by engaging with producers around the world. To date, more than 1,100 seafood farms are ASC-certified with a total production volume of 1.8 million tons.[17] As China is the largest aquaculture producer and consumer in the world, the ASC has made various efforts since its creation to introduce its standards to the country. As of October 2019, there are 17 ASC-certified farms in China (with two more in the process of being assessed) and 147 companies adopting the ASC CoC standard to handle certified products.[18]

The ASC was disseminated in China's seafood sector through both market and non-market channels. On the one hand, the demand of Western buyers, especially European ones, for ASC certification has led export-oriented Chinese producers to adopt the relevant standards. This mechanism is well demonstrated by the tilapia industry in Hainan province, which relies heavily on exports.[19] Like the MSC, the ASC has also participated in the Sustainable Seafood Forum and Sustainable Consumption Week discussed above. These activities have been helpful to raising awareness of buyers and consumers in China about sustainable aquaculture and the ASC certification, although as a younger scheme, it remains less known than the MSC.

On the other hand, being aware of the important role of the state in the market, the ASC has sought endorsement by government agencies and state-sponsored industry associations in China to promote itself. In this vein, the direct collaboration between the ASC and CAPPMA is critical. In fact, the ASC's initial entry into China in 2012–2014 began with a joint WWF and CAPPMA project, which was funded by the European Commission. Through this project, CAPPMA helped the ASC introduce its standards to a few large tilapia producers and coordinated with the regulatory agency on certification (i.e., CNCA) to facilitate the undertaking of audits.[20] As a result of the project, three tilapia farms obtained ASC certification in 2015. The project allowed CAPPMA and the ASC to better understand each other's interests and build good relationships. As a result, CAPPMA continued its support for the ASC after the project ended by disseminating information on the ASC in China's seafood industry. For instance, CAPPMA built connections between major Chinese producers and the ASC to initiate a new ASC standard for flatfish, a species produced and consumed mainly in East Asia and help gather experts and stakeholders for standard development.[21] In the case of the ASC, therefore, CAPPMA again has played a key role in the rise of private governance in China's aquaculture industry.

Best Aquaculture Practices

The Best Aquaculture Practices (BAP) is another certification program for farmed seafood, created by the Global Aquaculture Alliance (GAA) in the early 2000s. BAP standards focus on environmental protection, social responsibility, food safety, and animal welfare, and they are specified for different businesses along the supply chains including farms, feed mills, hatcheries, and processing plants. BAP began to rise in popularity in 2005 after gaining support from some large seafood brands and retailers, especially Walmart. As of October 2018, it had awarded certification to 1,451 farms, which produce over 1.5 million tons of seafood annually (BAP 2018b). In China today, there are 125 BAP-certified farms, 57 plants, 18 hatcheries, and 11 feed mills; most of these certified producers focus on the production of shrimp and tilapia.[22]

BAP was the first aquaculture certification scheme introduced to China back to the mid-2000s. Like the MSC, the scheme's uptake in China was initially driven by multinational retailers, especially Walmart, which in 2005 required all of its shrimp suppliers to be BAP-certified. Thus, for nearly a decade, certified producers in China centered around large, export-oriented companies that were capable of using industrial farming and vertically coordinating their supply chain, such as Zhanjiang Guolian, China's leading shrimp producing company.[23] This market dynamic also explains the concentration of Chinese certified producers in the shrimp and tilapia industry as China has been a major exporter of these two species to the US seafood market.

This pattern has begun to change over the last five years as the GAA has vigorously promoted BAP-certified products in China's domestic market. This change occurred due to the GAA's efforts to engage Chinese stakeholders. To promote BAP standards among more Chinese businesses, the GAA has not only gradually expanded its local team in China, but more importantly, it has also built partnerships through Memoranda of Understanding with industry associations (i.e., CAPPMA and China Entry Exit Inspection and Quarantine Association) and governmental agencies (CNCA) (Undercurrent News 2017). In particular, CAPPMA has provided critical support to many activities organized by the GAA in China for sharing information about and raising awareness of BAP certification.[24] With these domestic actors' endorsement, BAP has been able to reach a range of businesses along China's seafood supply chain including large retailers and seafood brands. As a result, BAP has won the support of several e-commerce platforms and become increasingly popular in the supply chain for China's domestic market (BAP 2017). A remarkable example is the commitment of JD.com, a major fresh produce e-retailer in China, to have 50% of its farmed seafood supply and over 80% of its private label farmed offerings as being BAP-certified by 2020 (BAP 2018a). All in all, with CAPPMA's support, BAP has proactively reached out to a large number of businesses in China's seafood supply chain, from farming companies to retailers, such that its certification has been quickly rising in prominence in the country.

Combining the insights drawn from the three certification schemes, we find that in the last fifteen years private seafood certification has gradually diffused throughout China. While buyers in developed countries initially drove Chinese producers to adopt the relevant standards, more recently these schemes originating in the West have enhanced their efforts to engage Chinese stakeholders and therefore increased their uptake in China's domestic market. In this process, domestic industry associations—especially CAPPMA—have played a central role in coordinating with public regulators and raising awareness of businesses. Figure 3 depicts stakeholder interactions for introducing and disseminating transnational certification in China's export and domestic

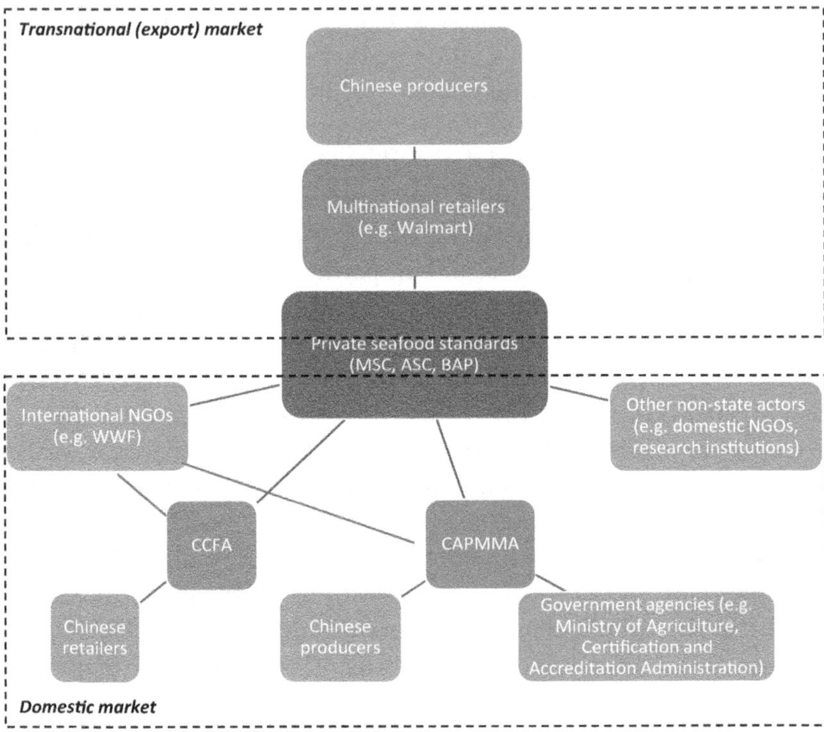

Fig. 3 Stakeholder interaction around transnational private governance in China

markets, respectively. The two markets have different pathways driving Chinese producers to support this new governance mode: one led by buyers' demand in foreign markets and the other based on engagement with domestic stakeholders. Due to growing consumption in China, the domestic market should have more sustainability impacts, so advocates of sustainable seafood should pay more attention to this pathway driven by domestic industry associations. Meanwhile, given China's institutionalized governance processes, NGOs play less prominent role in promoting private governance in China and tend to use non-confrontational strategies. Lastly, while some research has suggested that competition or cooperation may occur among certification schemes in the same issue domain (e.g., Auld 2014), we have not seen much interaction between ASC and GAA (or BAP) in terms of their activities in China.

Outputs and Outcomes of Transnational Private Governance in China

Bearing in mind different pathways for the rise of transnational private certification in China's seafood sector, I now assess the influence of this new governance mode on China's sustainability governance. The analysis here focuses on outputs and outcomes as the environmental impacts are difficult, and maybe too early, to be observed. Comparing the wild catch and aquaculture supply chains, I find that private certification has brought more changes in the latter, as many fish farms located in China adopted the relevant private standards in their production processes, especially those producing species targeting Western markets. In contrast, while the MSC has actively promoted certified products with some retailers, only one fishery in China's domestic waters is MSC-certified. Table 1 summarizes these variations across the end markets and subsectors. In short, while this study shows growing awareness of sustainability issues in China's seafood industry, significant barriers still exist for many Chinese stakeholders to adopt new practices.

Table 1 Level of influence of private certification in China's seafood industry

	Export market	Domestic market
Wild catch	Influential for processors but not much for fisheries	Almost no influence for fisheries, but quite influential for retailers
Aquaculture	Highly influential for producers	Influential for some producers and retailers

Outputs

Section "The Rise of Private Sustainability Certification in China's Seafood Sector" documented many efforts that have been made through certification to establish collaboration with domestic regulators and state-sponsored industry associations. In these processes of engagement, the relevant schemes and their partner NGOs have primarily relied on moral persuasion to raise awareness and generate interest of Chinese state actors. In line with the state's dominant role in China's environmental governance processes, this strategy contrasts with activist campaigns directly targeting large, branded companies, which have driven the rise of eco-certification in Western markets (e.g., Dauvergne 2017; Bloomfield 2017). Here we can see that transnational certification programs have attempted to leverage all six mechanisms of influence identified in Chapter "Introduction" (also Guttman et al. 2018) to motivate Chinese businesses for adopting sustainability standards.[25] While these activities so far seem effective in ensuring legitimacy of the relevant schemes in China and raising awareness of domestic stakeholders, most Chinese companies do not feel direct pressure in the market for the adoption of sustainability standards.

Another shortcoming of the existing strategy used by transnational certification schemes in China is the lack of direct engagement with upstream producers in the seafood supply chain, especially small-scale producers. To expand their influence in China's domestic market, certification schemes increasingly have invested their energy in gaining the support of downstream Chinese businesses and succeeded in convincing some retailers to make sourcing commitments. However, these organizations have paid insufficient attention to various challenges that many producers face to adopt their standards, including the lack of capacity and the poor fit of transnational rules with local contexts. According to a Chinese expert, certification programs and their supporters from the West generally do not consider local circumstances when introducing their standards to China and cannot provide sufficient assistance to producers for capacity building.[26] Consequently, transnational standards may remain too high or impractical for many Chinese producers such that certification would simply exclude them from the market. In this situation, transnational private governance may undermine sustainability of China's seafood industry. This finding echoes past research showing Southern producers' lack of involvement in transnational private governance, which is likely to perpetuate and even exacerbate inequalities in global value chains (e.g., Bennett 2017; Renckens and Auld 2019). Partly as a response to this problem, some Chinese NGOs and businesses have begun recently to develop their own initiatives for sustainability governance by building capability of fish farmers while taking into account local circumstances (Sun and van der Ven 2020).

Lastly, consumer education is also a key part of the work in China for many transnational private certification schemes. In this respect, we have seen more and more campaigns aiming to introduce the concept of sustainable seafood to Chinese consumers, especially among middle and upper middle class people in large cities (Fabinyi 2016). Although some surveys suggest that more and more Chinese consumers are concerned about the sustainability of the seafood sector (e.g., GoalBlue 2019), these educational activities may not lead to significant changes in consumption behavior in the market in the near future. A main reason is that most consumers remain unfamiliar with the relevant schemes and therefore cannot fully understand the meaning of sustainable seafood (Fabinyi et al. 2016). Furthermore, consumers' willingness to pay a premium for certified products may be too low to drive a transformation in the whole supply chain (Li et al. 2017; Fesenfeld et al. 2021). In this sense, transnational certification schemes have introduced a new concept and governance mode only to a very small group of Chinese consumers.

Outcomes

To measure the outcomes of private certification in China, I consider both ideational and behavioral changes of relevant stakeholders. In terms of ideational changes, a range of activities organized by private certification schemes, such as the Sustainable Seafood Forum, have increased Chinese stakeholders' awareness about sustainability issues associated with seafood production and consumption. This change has been observed by some foreign experts, who have noted an increase in the range of people participating in Forum discussions and in the number of Chinese companies interested in environmental and social issues.[27]

More specifically, some Chinese companies have changed their attitude on sustainability certification over the years, championing certification as a strategy for building brand reputation and expanding their market. As a result, they continue to invest in ways to improve compliance with certification standards. For example, a large tilapia producing company originally saw certification as an extra burden imposed by their foreign customers, but they recently began to use certification as a marketing strategy to promote their products in the domestic market.[28] However, with respect to domestic regulators, there is still no evidence showing formal support of government agencies (e.g., the Bureau of Fisheries in the Ministry of Agriculture or CNCA) for any private certification schemes.

In terms of behavioral change, certification schemes have been unfortunately unable to trigger a broad transformation in China's seafood industry. The explanation is threefold. First, knowledge gaps persist to prevent many actors along the supply chain from acting. As described by the founder of a local NGO, for China's seafood industry "knowing it (the concept of sustainability) does not necessarily mean also knowing how to do it, or when to do it" (cited from Godfrey 2017). Second, to date, only a very small segment of the industry has embraced private sustainability governance, and most adopters of transnational private standards are large producers having financial resources and technical capability and also aim to expand exports. This uptake pattern means that millions of small-scale producers supplying the domestic market are beyond the reach of transnational private governance and lack incentives to change their practices. Even for those that have been certified, their prior practices may be close to the relevant standards already, so that the additionality resulting from certification is small. As transnational certification programs can hardly reach out to these small-scale producers, they need to further collaborate with the Chinese government, especially agencies at the local level to make the relevant private standards more impactful.

Third, some existing public policies in China seem incompatible with practices championed by transnational certification schemes, constituting barriers to the adoption of the relevant standards. This phenomenon can be shown by the challenge of conducting MSC fishery certification in China. The fact that only one fishery in China's territorial waters is MSC-certified reveals a mismatch between existing fisheries practices in China and the management approach advocated by certification schemes developed in the West. To date, China's marine fisheries governance relies on blunt input control measures, especially seasonal fishing moratoria, instead of output controls like total allowable catch limits that require more resources to implement (Shen and Heino 2014). As a result, most Chinese fisheries lack detailed data, which is a minimum requirement for certification.[29] But this problem is too costly for individual fisheries to solve. Hence, the promise of transnational private governance in China is likely to depend on public policy changes. For instance, in 2018 a fishery of silver-stripe round herring in Zhejiang province began its assessment for MSC certification after the fishery had been identified by the central government as a pilot project of total allowable catch management.[30] This case suggests a need for better coordination and integration between public and private governance.

Conclusion

The case of transnational seafood certification exemplifies the institutionalized governance processes in China where the state has strong influence over the operation of non-state actors and governance initiatives. To disseminate their standards to and raise awareness of Chinese stakeholders, private certification schemes originating in the West have adapted to the Chinese context by relying on the strategy of engagement, instead of putting pressure on the government and businesses. By doing so, they have been able to leverage the influence of government-sponsored industry associations to change the attitude of some producers and supply chain actors and ultimately generate momentum in the uptake of their standards. In other words, the support of these associations, which enhance the legitimacy of private governance in the eyes of Chinese stakeholders, is indispensable to the recent rise of seafood certification in China.

Nonetheless, this chapter shows that in many aspects the influence of private certification on the sustainability of China's seafood sector remains limited. At the level of outputs, transnational private governance schemes have yet to actively involve upstream producers in China and the effects of their consumer education activities remain far from evident. At the level of outcomes, despite growing awareness of many stakeholders about the concept of sustainability, a large number of Chinese producers face significant challenges to adopt relevant standards. To address them, some support from the Chinese state, including capacity building and policy reforms, may be needed.

Looking ahead, private certification schemes could achieve greater impact in promoting sustainable seafood production and consumption in China in the coming decade if they strengthen their collaboration with local industry associations. In this scenario, producers and supply chain actors like retailers would pay more attention to environmental and social impacts of their businesses, while the awareness of consumers of sustainable seafood continues to increase. This could constitute a virtuous cycle to reform China's seafood sector, which would make important contributions to sustainability of global seafood food value chains. Moreover, if China's seafood industry further embraces transnational certification schemes, such support for private governance can change market dynamics and influence stakeholders in other emerging economies including producers supplying the growing Chinese market as well as producers competing with Chinese producers to export to Western markets. This would provide momentum for the expansion of existing, Western-developed certification schemes. However, as a marked-based voluntary instrument for corporate social responsibility, certification seems unlikely to drive a fundamental transformation in the whole industry, unless the Chinese government provides further support to certification such as setting requirements to make relevant standards de facto mandatory.

The case of seafood certification in China shows that non-state actors, including both NGOs and businesses, cannot supersede the state completely in supplying sustainability governance. Instead, the state remains highly relevant to and influential in transnational private governance. In other words, to make the seafood supply chain of the world's largest producing and consuming country more sustainable, it is necessary to strategically engage the Chinese state and proactively help domestic regulators make reforms. This chapter thus confirms the proposition of some experts on seafood governance that "states will ultimately determine the success or failure of the sustainable seafood movement" (Bush and Oosterveer 2019: 197). This insight is in line with evidence from some other emerging economies where the state has intervened to establish sustainability governance interacting with private certification (Sun and van der Ven 2020). While it is hard to predict what actions the Chinese state will take toward private governance in the future, we must recognize that in China, as well as in other parts of the non-Western world, the state cannot be bypassed in governing fisheries and seafood.

Notes

1. The notion of governance distinguishes networks or organizations that only influence the creation and operation of other institutions but are not recognized as authoritative, and those that set their own rules for others to comply with. The focus of this chapter is these rule-making governance systems led by non-state actors, which encompass the second and third streams of activities of non-state actors in environmental governance identified by Guttman et al. (2018).
2. The relevant terminology includes "non-state market-driven governance," "transnational private governance," "transnational private regulation," and "transnational business governance."
3. Remarkable exception regarding the work on eco-labels in China includes (Bartley 2018; Schleifer and Sun 2018; Sun 2018, 2022).
4. This chapter focuses on the adoption and implementation of standards, instead of monitoring and verification. For this reason, auditors are not included.
5. It is not suggested here that NGOs adopt non-confrontational strategies only in China, but compared to Western democracies, NGOs in China rely on confrontational action like boycotts much less often.
6. The difference between *Shiye danwei* and *Shehui tuanti* is worth noting. I focus on the former in my study as they function more like an extension of the Chinese bureaucracy. But in 2015, the Chinese government issued a

plan on reform of industry and business associations by encouraging them to detach from the government agency supervising them. As a result of this reform, some of these associations began the process to change their status and gradually became *Shehui tuanti*, which are less dependent on the government.

7. "Seafood" here includes the fish and fishery products from seawater as well as freshwater.
8. While different rules exist at the international and domestic levels to regulate each subsector, the Ministry of Agriculture is the Chinese state agency responsible for both subsectors.
9. Data drawn from the MSC website in October 2019.
10. Interview with the MSC China representative, Beijing, July 2015.
11. Interview with an official in the MSC's headquarters, London, April 2016.
12. Interview with the general manager of a leading certifier for the MSC in China, Qingdao, August 2016.
13. Interview with the MSC program director in China, Beijing, July 2015.
14. Interview with the MSC global business director, London, April 12, 2016.
15. See the introduction on the association's webpage: http://en.cappma.org/about%20cappma.html. Interview with an official of the association, Beijing, March 21, 2017.
16. Interview with a former official working for the fisheries program in WWF-China, Qingdao, April 2017.
17. Data from the ASC's website, accessed at https://mailchi.mp/0384c0fdb64e/xr162vrjvq-2434961. The proportions of the ASC-certified products over the global total production vary across species: 2% for Bivalve, 13% for Pangasius, 29% for Salmon 49% for Shrimp, 3% for Tilapia, and 4% for Trout.
18. Data drawn from the ASC's website by searching certified farms and suppliers in China, http://asc.force.com/Certificates/ and https://www.asc-aqua.org/what-you-can-do/take-action/find-a-supplier/.
19. Interview with officials of a local NGO promoting sustainable seafood, Haikou, April 2017.
20. In China, the third-party auditors for each new standard must be approved by the CNCA. For the ASC certification in this project, the process was streamlined thanks to CAPPMA's support (interview with a former official of WWF-China, Qingdao, April 2017).
21. Interview with the ASC China representative, Qingdao, April 2017.
22. Data from an online database of certified producers on the BAP's website, accessed at https://bapcertification.org/Producers.

23. Interview with GAA Asia Business Development Director, Shanghai, March 2017.
24. Interview with GAA Asia Business Development Director, Shanghai, March 2017.
25. These mechanisms are: the shadow of the state, crisis, provision of resources essential to enterprise activity, provision of expertise, push and pull of going abroad, and cross border cultural arbitrage.
26. Interview with officials of a Chinese NGO promoting sustainable seafood, Haikou, April 2017.
27. Phone interview with a senior official of a transnational certification scheme, September 2017.
28. Interview with the general manager of a tilapia producing company, Wenchang, April 2017.
29. Interview with an MSC auditor in China, Qingdao, April 2017.
30. Information disseminated by the MSC through its newsletter in China, the author's personal communication, December 2018.

References

Andonova, Liliana B., Michele M. Betsill, and Harriet Bulkeley. 2009. Transnational Climate Governance. *Global Environmental Politics* 9 (2): 52–73. https://doi.org/10.1162/glep.2009.9.2.52.

Auld, Graeme. 2014. *Constructing Private Governance: The Rise and Evolution of Forest, Coffee, and Fisheries Certification.* New Haven: Yale University Press.

BAP. 2017. GAA, BAP Make Splash at China Seafood & Fisheries Expo. Best Aquaculture Practices—Best Aquaculture Practices Certification. https://bapcertification.org/blog/gaa-bap-china-expo/.

BAP. 2018a. China's Largest Retailer Commits to BAP for More Responsible Seafood. Best Aquaculture Practices—Best Aquaculture Practices Certification. https://bapcertification.org/blog/jdcom-new-commitment/.

BAP. 2018b. Number of BAP-Certified Facilities Hits 2,155. Best Aquaculture Practices—Best Aquaculture Practices Certification. https://bapcertification.org/blog/2155-bap-certified-facilities/. Accessed October 22, 2019.

Bartley, Tim. 2007. Institutional Emergence in an Era of Globalization: The Rise of Transnational Private Regulation of Labor and Environmental

Conditions. *American Journal of Sociology* 113 (2): 297–351. https://doi.org/10.1086/518871.

Bartley, Tim. 2018. *Rules Without Rights: Land, Labor, and Private Authority in the Global Economy*. Transformations in Governance. New York, NY: Oxford University Press.

Bennett, Elizabeth A. 2017. Who Governs Socially-Oriented Voluntary Sustainability Standards? Not the Producers of Certified Products. *World Development* 91 (March): 53–69. https://doi.org/10.1016/j.worlddev.2016.10.010.

Bernstein, Steven, and Benjamin Cashore. 2007. Can Non-state Global Governance Be Legitimate? An Analytical Framework. *Regulation & Governance* 1 (4): 347–371. https://doi.org/10.1111/j.1748-5991.2007.00021.x.

Bloomfield, Michael John. 2017. *Dirty Gold: How Activism Transformed the Jewelry Industry*. Earth System Governance. Cambridge, MA: MIT Press.

Bush, Simon R., and Peter Oosterveer. 2019. *Governing Sustainable Seafood*. Earthscan Food and Agriculture Series. London and New York: Routledge and Taylor & Francis Group.

Cao, Ling, Yong Chen, Shuanglin Dong, Arthur Hanson, Bo Huang, Duncan Leadbitter, David C. Little, et al. 2017. Opportunity for Marine Fisheries Reform in China. *Proceedings of the National Academy of Sciences* 114 (3): 435–442. https://doi.org/10.1073/pnas.1616583114.

Cashore, Benjamin. 2002. Legitimacy and the Privatization of Environmental Governance: How Non-state Market-Driven (NSMD) Governance Systems Gain Rule-Making Authority. *Governance* 15 (4): 503–529. https://doi.org/10.1111/1468-0491.00199.

Cashore, Benjamin William, Graeme Auld, and Deanna Newsom. 2004. *Governing Through Markets: Forest Certification and the Emergence of Non-state Authority*. New Haven: Yale University Press.

Cui, He. 2015. China's Fisheries in Transition (Zhuang xing zhong de zhong guo yu ye). *China Fisheries News*.

Dauvergne, Peter. 2017. Is the Power of Brand-Focused Activism Rising? The Case of Tropical Deforestation. *The Journal of Environment & Development* 26 (2): 135–155. https://doi.org/10.1177/1070496517701249.

Fabinyi, Michael. 2016. Sustainable Seafood Consumption in China. *Marine Policy* 74 (December): 85–87. https://doi.org/10.1016/j.marpol.2016.09.020.

Fabinyi, Michael, Neng Liu, Qingyu Song, and Ruyi Li. 2016. Aquatic Product Consumption Patterns and Perceptions Among the Chinese Middle Class.

Regional Studies in Marine Science 7 (September): 1–9. https://doi.org/10.1016/j.rsma.2016.01.013.

FAO, ed. 2016. *The State of World Fisheries and Aquaculture 2016: Contributing to Food Security and Nutrition for All*. The State of World Fisheries and Aquaculture 2016. Rome.

Fesenfeld, L. P., Sun, Y., Wicki, M., & Bernauer, T. 2021. The role and limits of strategic framing for promoting sustainable consumption and policy. Global Environmental Change, 68, 102266. https://doi.org/10.1016/j.gloenvcha.2021.102266.

GoalBlue. 2019. *First-Tier Chinese Cities Sustainable Fishery Consumption Behavior Survey Report*. Shenzhen: GoalBlue.

Godfrey, Mark. 2017. Q&A with Han Han, Founder and CEO of China Blue. Seafood Source. https://www.seafoodsource.com/news/environment-sustainability/q-a-with-han-han-founder-and-ceo-of-china-blue.

Gulbrandsen, Lars H. 2009. The Emergence and Effectiveness of the Marine Stewardship Council. *Marine Policy* 33 (4): 654–660. https://doi.org/10.1016/j.marpol.2009.01.002.

Gulbrandsen, Lars H. 2014. Dynamic Governance Interactions: Evolutionary Effects of State Responses to Non-state Certification Programs: Dynamic Governance Interactions. *Regulation & Governance* 8 (1): 74–92. https://doi.org/10.1111/rego.12005.

Guttman, Dan, Oran Young, Yijia Jing, Barbara Bramble, Bu Maoliang, Carmen Chen, Kathinka Furst, et al. 2018. Environmental Governance in China: Interactions Between the State and 'Nonstate Actors'. *Journal of Environmental Management* 220: 126–135. https://doi.org/10.1016/j.jenvman.2018.04.104.

Hale, Thomas, and David Held (eds.). 2011. *Handbook of Transnational Governance: Institutions and Innovations*. Cambridge and Malden MA: Polity.

Ho, Peter. 2007. Self-Imposed Censorship and De-politicized Politics in China: Green Activism or a Color Revolution? In *China's Embedded Activism: Opportunities and Constraints of a Social Movement*, edited by Peter Ho and Richard Louis Edmonds, 20–43. London: Routledge.

Jacquet, Jennifer, John Hocevar, Sherman Lai, Patricia Majluf, Nathan Pelletier, Tony Pitcher, Enric Sala, Rashid Sumaila, and Daniel Pauly. 2010. Conserving Wild Fish in a Sea of Market-Based Efforts. *Oryx* 44 (01): 45. https://doi.org/10.1017/S0030605309990470.

Li, Yan, Lei Zhang, and Min Jin. 2017. Report on Consumer Awareness and Behaviour Change in Sustainable Consumption. China Sustainable

Consumption Research Program. The 10-Year Framework of Programme on Sustainable Consumption and Production Patterns.

Lister, Jane. 2011. *Corporate Social Responsibility and the State: International Approaches to Forest Co-regulation.* Vancouver: UBC Press.

Mertha, Andrew. 2009. 'Fragmented Authoritarianism 2.0': Political Pluralization in the Chinese Policy Process. *The China Quarterly* 200 (December): 995. https://doi.org/10.1017/S0305741009990592.

MSC. 2017. MSC and Chinese Partners Including Tmall Announce Commitment to Sustainability | Marine Stewardship Council. June 9, 2017. https://www.msc.org/media-centre/press-releases/msc-and-chinese-partners-including-tmall-announce-commitment-to-sustainability.

Pikitch, E.K. 2012. The Risks of Overfishing. *Science* 338 (6106): 474–475.

Potts, Jason, Ann Wilkings, Matthew Lynch, and Scott McFatridge. 2016. *State of Sustainability Initiatives Review: Standards and the Blue Economy.* Winnipeg: International Institute for Sustainable Development. http://publiceblib.com/choice/publicfullrecord.aspx?p=4532673.

Renckens, Stefan, and Graeme Auld. 2019. Structure, Path Dependence, and Adaptation: North-South Imbalances in Transnational Private Fisheries Governance. *Ecological Economics* 166 (December): https://doi.org/10.1016/j.ecolecon.2019.106422.

Schleifer, Philip, and Yixian Sun. 2018. Emerging Markets and Private Governance: The Political Economy of Sustainable Palm Oil in China and India. *Review of International Political Economy* 25 (2): 190–214. https://doi.org/10.1080/09692290.2017.1418759.

Schouten, Greetje, and Verena Bitzer. 2015. The Emergence of Southern Standards in Agricultural Value Chains: A New Trend in Sustainability Governance? *Ecological Economics* 120 (Supplement C): 175–184. https://doi.org/10.1016/j.ecolecon.2015.10.017.

Shen, Gongming, and Mikko Heino. 2014. An Overview of Marine Fisheries Management in China. *Marine Policy* 44 (February): 265–272. https://doi.org/10.1016/j.marpol.2013.09.012.

Sun, Yixian. 2018. *Beyond the North: The Diffusion of Private Sustainability Governance in China.* Geneva: Graduate Institute of International and Development Studies (IHEID).

Sun, Yixian. 2022. Certifying China: The Rise and Limits of Transnational Sustainability Governance in Emerging Economies. MIT Press.

Sun, Yixian, and Hamish van der Ven. 2020. Swimming in Their Own Direction: Explaining Domestic Variation in Homegrown Sustainability

Governance for Aquaculture in Asia. *Ecological Economics* 167 (January): 106445. https://doi.org/10.1016/j.ecolecon.2019.106445.

Teets, Jessica. 2018. The Power of Policy Networks in Authoritarian Regimes: Changing Environmental Policy in China. *Governance* 31 (1): 125–141. https://doi.org/10.1111/gove.12280.

Teets, Jessica C. 2014. *Civil Society Under Authoritarianism: The China Model*. New York: Cambridge University Press.

Undercurrent News. 2017. GAA Expands China Presence with CAPPMA Partnership, New Office. January 12, 2017. https://www.undercurrentnews.com/2017/01/12/gaa-expands-china-presence-with-cappma-partnership-new-office/.

van der Ven, H., Sun, Y., & Cashore, B. 2021. Sustainable commodity governance and the global south. *Ecological Economics*, 186, 107062. https://doi.org/10.1016/j.ecolecon.2021.107062.

van Rooij, Benjamin, Rachel E. Stern, and Kathinka Fürst. 2016. The Authoritarian Logic of Regulatory Pluralism: Understanding China's New Environmental Actors. *Regulation & Governance* 10 (1): 3–13. https://doi.org/10.1111/rego.12074.

Walmart. 2006. Wal-Mart Takes Lead on Supporting Sustainable Fisheries. February 3, 2006. https://corporate.walmart.com/_news_/news-archive/2006/02/06/wal-mart-takes-lead-on-supporting-sustainable-fisheries.

Wijaya, A., and P. Glasbergen. 2016. Toward a New Scenario in Agricultural Sustainability Certification? The Response of the Indonesian National Government to Private Certification. *The Journal of Environment & Development* 25 (2): 219–246. https://doi.org/10.1177/1070496516640857.

Young, Oran R., ed. 1999. *The Effectiveness of International Environmental Regimes: Causal Connections and Behavioral Mechanisms*. Global Environmental Accord. Cambridge, MA: MIT Press.

Young, Oran R. 2011. Effectiveness of International Environmental Regimes: Existing Knowledge, Cutting-Edge Themes, and Research Strategies. *Proceedings of the National Academy of Sciences* 108 (50): 19853–19860.

Young, Oran R. 2013. *On Environmental Governance: Sustainability, Efficiency, and Equity*. On Politics. Boulder: Paradigm Publishers.

Young, Oran R., Dan Guttman, Ye Qi, Kris Bachus, David Belis, Hongguang Cheng, Alvin Lin et al. 2015. Institutionalized Governance Processes: Comparing Environmental Problem Solving in China and the United States. *Global Environmental Change* 31 (March): 163–173. https://doi.org/10.1016/j.gloenvcha.2015.01.010.

Green Supply Chain Initiatives in China: The Roles of Nonstate Actors

Lingxuan Liu

Introduction

This chapter provides a preliminary account of several case studies in environmental governance being developed to explore non-regulatory efforts to tackle environmental challenges associated with global supply chains in China. In doing so, the chapter applies concepts and insights from both environmental governance and supply chain management. With a focus on the green supply chain initiative (GSCI) that China introduced in 2014, the chapter examines four pioneering cases involving Chinese cities and a cohort of manufacturing firms that have become the pioneers of GSCIs. The chapter begins with an analytical framework to explore the design and implementation of the initiatives. This exploration aims to generate new understandings of the roles of nonstate actors

L. Liu (✉)
Pentland Centre for Sustainability in Business, Lancaster University, Lancaster, UK
e-mail: lingxuan.liu@lancaster.ac.uk

D. Guttman et al. (eds.), *Non-state Actors in China and Global Environmental Governance*, Governing China in the 21st Century,
https://doi.org/10.1007/978-981-33-6594-0_10

in environmental governance as well as the specific nonstate mechanisms of influencing environmental practices in both the public sector and the private sector.

Environmental governance in a global context has been evolving continuously during the past several decades. The efforts include global environmental governance "regimes" (e.g. the Montreal Protocol, the climate regime) and regional environmental governance systems (e.g. the regime for the Greater Mekong)[1] (Young et al. 2015). Such regimes are based on and defined by agreements among sovereign states. Nonetheless, in the evolution and implementation of such hierarchical structures, nonstate actors play increasingly critical roles (Teegen et al. 2004; Cashore et al. 2004; Arts 2006; Doh and Guay 2006). For example, the environmental performance of the global manufacturing and consumption systems is essential to sustainability, yet most industrial and consumer products now have extensive global supply chains that are often beyond the control of individual states.

China's remarkable economic development over the past four decades has been guided by top-down strategies developed and implemented by the central government. This hierarchical structure also has characterised the emergence of post-1970s environmental governance. Such approach features Five-Year Plans (FYPs) with increasingly demanding environmental objectives, top-down political campaigns (Tanner 2005), and the promulgation of laws and regulations. The same approach also focuses on the development of effective compliance mechanisms at local level with limited budgets and resources (Young et al. 2015; Liu et al. 2015). Meanwhile, China also has a growing awareness of the importance of supplementing "top-down" environmental governance with efforts by nonstate actors, a development that provides the context for this chapter.

As the focus of China's leadership on the environment has intensified, central government efforts have highlighted the need for a complementary perspective to address the complexities of sustainable development, particularly in an economy that are globally interconnected through trade. In 2017, the activities of central environmental inspection teams[2] led by China's Ministry of Environmental Protection (MEP) (since restructured as the Ministry of Ecology and Environment or MEE) have possibly led to business disturbance and shutdowns, causing economic disruptions and supply chain instability in many industries.[3] In a separate development starting in 2017, China decided to rush to embrace natural gas as a cleaner energy alternative to coal. It did so with a similar reliance on a top-down administrative approach in both manufacturing and residential sectors, ordering the removal of coal-burning equipment and its replacement with gas-based equipment. The campaign led to a significant shortage of natural gas in China and a sharp increase in the import price of gas.[4]

In both cases, these state-directed actions have failed to foresee, or at least have chosen to neglect, the negative impacts and chain reactions in the business world, reflecting the limitation of decision-making on the part of state actors. This chapter argues that environmental governance studies must focus not only on the role of state and nonstate actors in domestic governance, but also on the global social, political or economic elements of environmental governance (Dallas et al. 2017).

Green Supply Chain Policies in China: A Context in the World's Factory

The Chinese manufacturing industries, including state- and nonstate-owned enterprises, produce more than 60% of the world's large electronic appliances, computers, mobile phones and various clothing products.[5] From the 1980s to the 2000s, most of China's manufacturing industries were export-oriented. Buyers from western countries began to realise that the outsourcing of manufacturing to China comes with the outsourcing of pollution, but few were motivated to manage the environmental problems caused by their supply chains. Most buyers insisted that their companies were responsible and their products were not harmful as long as their suppliers were compliant with local regulations, which usually have lower environmental standards. During that period, the idea of "green supply chains" was seldom mentioned as a policy-related term.

Since the 1990s, China has come to focus on "green" policies and laws. In the early twenty-first century, for example, these efforts included the 2008 circular economy law, the initiation of numerous local experiments in "eco-cities"[6] and "ecological industrial parks," and the promotion of "low carbon development." Most of these concepts and terms came from developed countries in Europe or North America, but they were adapted to underpin political campaigns in the regime of China's multi-level governance and political incentives (Liu et al. 2013). Those experiments were, in turn, part of China's longer and broader use of local "experiments" or "pilot projects" as a means of economic development as well as institutional innovation (Liu et al. 2013).

The current activities that are the focus of this chapter occur at the nexus of two developments: (i) China's extensive use of local experiments as a means of development and (ii) China's growing awareness of the importance of green supply chains in the broader context of the now decade-long commitment of China's leaders to the development of "ecological civilization."[7]

The concept of a Green Supply Chain (GSC) was first mentioned in China in 1999 by the Ministry of Commerce in connection with a government project called "Triple Green." The project focused on food security issues (including "green food") and sought to promote "green consumption, green market and green channels" (explained with green logistics and supply chains). The slogans soon became part of the political rhetoric of a "green economy" in China, which came to be applied to every industry or sector, often with limited concrete detail as to what "green" meant. The term "green economy" was then replaced by new political slogans, such as the Circular Economy.

In 2008, the Environmental Defence Fund (EDF), a US-based NGO with a long-standing presence in China, initiated a pilot Green Supply Chain project with Walmart.[8] The idea was to implement critical lessons from business cases in North America, based on an Operations Management concept developed at Michigan State University (Handfield et al. 1997), in order to improve the environmental performance of manufacturing industries in China. The idea caught the attention of the China Council for International Cooperation on Environment and Development (CCICED),[9] a high-level government organised non-governmental organisation (GO-NGO) with a commitment to produce joint Chinese/foreign expert reports presented each autumn to China's Premier. When CCICED proposed the idea of green supply chain management to the State Council in 2011, the State Council was positive about it.

Since 2014, the central government has published a series of policies, plans and announcements (Table 1) that encourage green supply chain initiatives. These government documents discuss green manufacturing, including resource efficiency, industrial and economic transitions, and technological innovation as well as environmental protection. The promotion of green supply chain management was much like previous political campaigns. The real puzzle was: *why promote GSCs as a new round of campaigns, and why now?*

To answer the question, a content analysis was performed examining recent (post-2014) critical national policy documents that mention the term GSC, whether as direct encouragement of GSC initiatives or as more indirect, contextual discussions. This analysis indicates that the interest of China's central government in promoting GSC stems from three sources.

1. A focus on GSCs is essential to achieve economic transformation through industrial innovation (Fig. 1 and the core agenda in Table 1). The national campaign known as "Made-in-China 2025" aims to improve the efficiency, innovation capacity and productivity of the manufacturing industries, initially up to 2025 and then on to 2035 and 2049. The goal is to transform China from a low value-added high pollution economy (e.g. industries such as steel, cement production or textile manufacturing) to a low pollution, innovative and more profitable economy. This campaign began in 2015 with 50 pilot projects across China, working with leading Chinese companies, such as Haier (air conditioners), BYD (electric cars) and DJI (drones). The goal of focusing on the high end of the value chain raises at least two challenges: Are the companies that become first movers ready to lead to a sustainable future? Will China gradually lose competitive advantages over emerging economies such as India or countries in Africa? In this context, GSC efforts may be important in finding solutions to these challenges.

The left side is documents proposed and implemented through the Ministry of Environmental Protection (MEP), while the right side is by the Ministry of Industry and Information Technology (MIIT). The lower

Table 1 National policies and plans that encourage green supply chain concepts

Policy or planning document	Governmental organisation of authority	Date of publication	Category of document	
Core agenda				
Made in China 2025	State Council	March 2015	National strategy	N/A
The 13th Five-Year Plan for economic and social development	NDRC	March 2016	Planning	*Gui hua*
The 13th Five-Year Plan for ecological and environmental protection	MEP	Dec 2016	Planning	*Gui hua*
Industrial green development plan, 2016–2020	MIIT	June 2016	Planning	*Gui hua*
Guidelines for implementing green manufacturing, 2016–2020	MIIT	Sept 2016	Guidelines	*Zhi nan*
Capacity building of green manufacturing system	MIIT	Sept 2016	Red headline	*Tong zhi*
Guidelines for establishing green manufacturing standards	MIIT	Sept 2016	Guidelines	*Zhi nan*
Action plan for promoting green manufacturing project	MEP	Dec 2016	Action plan	*Fang'an*

(continued)

Table 1 (continued)

Policy or planning document	Governmental organisation of authority	Date of publication	Category of document	
Action plan of industrial energy-saving and green standardisation, 2017-2019	MIIT	May 2017	Action plan	*Fang'an*
Contextual agenda				
Announcement at the 22nd APEC leadership meeting	–	Nov 2014	Announcement	N/A
Opinions of promoting supply chain innovation and implication	State Council	Oct 2017	Opinions	*Yi jian*
Action plan of controlling GHG emissions, 13th FYP	State Council	Oct 2016	Action plan	*Fang'an*
Action plan of environmental top-runner	MEP	July 2015	Action plan	*Fang'an*
Opinions on promoting green lifestyle	MEP	Dec 2016	Opinion	*Yi jian*
Guidelines of corporate green procurement (pilot version)	Ministry of Commerce, MEP, MIIT	Dec 2016	Guidelines	*Zhi nan*
Guidelines for implementing green "Belt and Road Initiative"	MEP	April 2017	Guidelines	*Zhi nan*

Note National Development & Reform Committee (NDRC)
Ministry of Environmental Protection (MEP)
Ministry of Industry and Information Technology (MIIT)

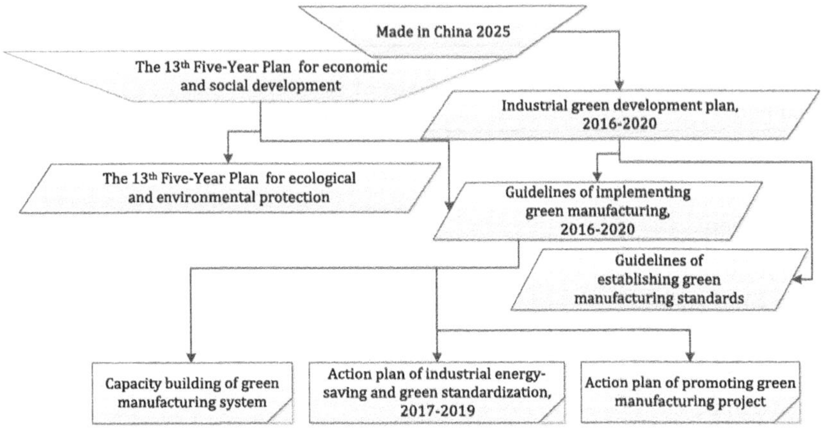

Fig. 1 The family of GSC-related policy documents at central level

tier policies, plans or guidelines were developed to support or implement the higher tier ones.

2. A GSC approach is essential to provide alternative governance strategies for environmental challenges (see Fig. 1 and the core agenda in Table 1). It has long been known that environmental enforcement in China at the local level can be problematic. In the context of environmental protection, the influence of "*difang baohu*" or "local protectionism" has lead to local actions that favour the growth of GDP over environmental protection. It does not necessarily mean a lack of concern for the environment, but rather the lack of resources to address economic and environmental goals simultaneously. Environmental non-compliance on the part of manufacturing industries can be explained as an effort to protect the local economy either directly or through *rent-seeking*, while even higher levels of compliance can generate "hotspots" of high environmental impacts due to the intensity of specific industries. The 2014 amendments to the Environmental Protection Law provided further means for enforcement, notably by allowing qualified NGOs to bring lawsuits against polluters on behalf of the public interest. Nonetheless, environmental compliance may well still require the installation of costly equipment or lead to a loss of productivity eliminating local jobs and reducing GDP. In this context, GSC efforts can draw on business and operations management insights and may offer new knowledge, resources and incentives to solve environmental challenges in the manufacturing sector.

3. A focus on GSC is essential to the Belt and Road Initiative (see contextual agenda in Table 1). China's Belt and Road Initiative (BRI) is the leadership's core global economic vision. The initiative has attained broad global recognition. At the same time, in prior and ongoing overseas activities, Chinese companies are increasingly being accused of causing environmental damage and extracting valuable local resources without adequate compensation or precaution. As the BRI gains momentum, China's development practices are gaining increased attention on the part of global and local NGOs and other local groups. Such attention includes a concern for environmental sustainability, but also for labour conditions and other considerations, such as long-term debt obligations.[10] Previous strategies and mindsets that had enabled overseas expansion now require critical attention as China's efforts become the focus of increasing global attention. In this context, previous efforts at GSC management, including multi-stakeholder processes, are a source of lessons the BRI is beginning to apply. For example, in May 2017, The Belt and Road Green Supply Chain Network (BR-GSCN) was founded in Beijing to facilitate sustainability collaborations among BRI regions.

Although the focal/buyer companies and suppliers are not state actors, for the clarity of discussions in this chapter, they are categorised as "business" instead of nonstate actors—since in current GSCM studies, buyers and suppliers are considered as core roles, while other stakeholders (state or nonstate) are external to the supply chain.

A consideration of those three types of central government motivation associated with green supply chain initiatives can help us understand the development of a complex landscape of environmental responsibilities that cut across geographical boundaries and regional jurisdictions. At the central level, the Ministry of Environmental Protection (MEP) and the Ministry of Industry and Information Technology (MIIT) have been keen to lead such initiatives. Their leadership might generate ambiguity, if not confusion, when it comes to local implementation. But this is certainly not the first time that multiple Chinese ministries with potential conflicts have become involved in a given environmental issue. That

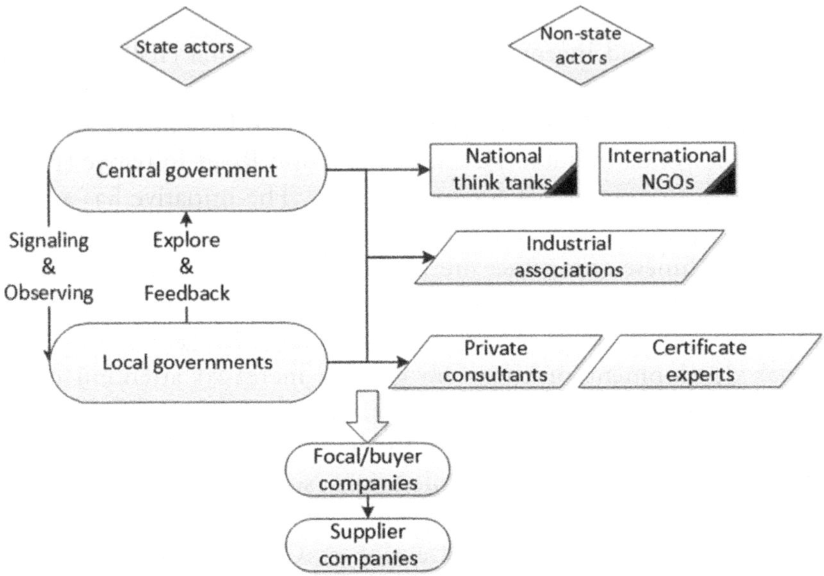

Fig. 2 A conceptual model of the network of state and nonstate actors in GSC collaboration

is probably why GCSI is not a clear and coherent policy initiative at local level, but rather a collection of actions that share some common elements. Because of this diversity and complexity, nonstate actors have a space to operate regarding supply chain management. Figure 2 provides a conceptual model of GSCI collaboration. The nonstate actors in this case include: buyers, suppliers, industrial associations, NGOs, think tanks and individual experts.

From Encouragement to Initiatives: A Tale of Four Cities

While the central government has sent "positive signals" regarding GSCIs, it is also evident that most local jurisdictions are not able to engage in regulation across the geographical boundaries of supply chains. So, how have specific GSCIs been evolving through central-local interactions and collaborations among state and nonstate actors?

In the document "*Guidance on the positive role of environmental protection in promoting supply-side structural reform*" published in April 2016, the MEP declared an intention "... *to encourage local governments to learn from the experience of* ***Shanghai, Tianjin, Shenzhen and Dongguan****, to promote GSCI among industries and industrial parks that have high volumes of pollutions, extensive supply chains, and great potential for green transition, to mobilise the leading roles of focal companies and key enterprises, and to organise and promote GSCI pilots for environmental management.*"

The policy documents do not provide a clear explanation for the choice of four cities as GSCI pilots. However, each case did receive "a supporting signal" from a central government source, and the city government in each case responded.

At the forefront, there is socio-economic common ground among the four GSCI pilot cities. They belong to developed regions of China that have (i) a solid foundation of manufacturing industries and related innovation capacity, (ii) access to knowledge and resources of business and NGOs at national and international levels and (iii) access to focal companies involved in multiple supply chains that may permit the influence of green manufacturing either upstream or downstream. Table 2 summarises key features of the four cities.

Because all the resultant initiatives are still within their first five-year period, analysis of their effectiveness is not yet possible. Instead, the

Table 2 Basic socio-economic features of the GSCI pilot cities

City	Location	Land area km^2	Population estimated in 2017	GDP in 2016 (billion CNY)
Tianjin	North China	11,760	12 million	1788.5
Shanghai	East China	6,340	24 million	2746.6
Shenzhen	South China	2,050	12 million	1949.2
Dongguan	South China	2,465	8 million	677.0

emphasis here will be on the design and formation of the initiatives and on the role of nonstate actors in each case.

Box 1 Relevant Socio-Economic Context of the Four Cities

Tianjin, like Shanghai, Beijing and Chongqing, is a metropolitan city operating directly under the central government (i.e. with province-like status) and is one of the earliest of China's industrialised cities. Tianjin University (created as Beiyang University in 1895) was the first university established in modern China. From as early as the 1920s, the city became a cradle of China's modern manufacturing industries. For example, the first bicycle and the first TV set in China were both produced in Tianjin. In 1950, the Tianjin-based "Flying Pigeon" Bicycle Company laid the foundation for China's bicycle culture. Tianjin is also the only Chinese city that has developed both the aviation industry and the aerospace industry. In November 2013, Tianjin published an *Action Plan of Green Supply Chain Pilot*, China's first local action plan for a GSCI.

In the first part of the twentieth century, Shanghai, noted for its foreign concessions, was a core global entrepôt to China. Following Deng Xiaoping's strategy in the 1990s to develop the Pudong District, Shanghai rapidly (re)assumed its role as an international metropolis with a full landscape of global businesses. For example, Shanghai hosts the regional headquarters of 572 multinational companies operating in China, 54 of which are Asia and Pacific headquarters offices. In the past decades, the core of Shanghai's economic growth has shifted from manufacturing industries (secondary sector) to business and service industries (tertiary sector).

Shenzhen, the most successful of the first Special Economic Zones created at the onset of China's Opening-up & Reform (*Gaige Kaifang*) is considered the southern centre of China's high-tech, IT, and electronics industries in terms of research and development as well as production. Famously, starting from nothing but rice paddies several decades ago, Shenzhen now ranks third (behind Beijing and Shanghai) in urban GDP. During a visit in 2019, Sebastian Piñera, the President of Chile commented: "Is there anything that Shenzhen cannot produce?"

Dongguan (located in Guangdong province, South China) has a history of export-oriented manufacturing industries starting from the opening up and reform of the 1980s. In 2014, two-thirds of Dongguan's GDP involved export-oriented manufacturing. Enterprises in Dongguan are supplying multinational companies, such as Walmart, Apple, Dell, IBM and Nike. In 2015, Dongguan became the only city formally approved by the Ministry of Environmental Protection as a GSCI pilot.

Central Signals

As indicated, two central government ministries encouraged the concept of green supply chain management. Due to their administrative responsibilities and established channels, MEP targeted city government agencies to encourage policymaking and the planning of GSCI, while MIIT targeted companies and factories to promote resource efficiency, technology innovation and transparency.

As shown in Table 3, two key nonstate actors have facilitated MEP's work: the China Council for International Cooperation on Environment and Development (CCICED) and the China—ASEAN Environmental Cooperation Centre (CAECC). Both organisations may be called GONGOs (government created organisations that have legal status of NGOs). The two organisations assumed significant responsibilities regarding international environmental collaboration since 1992 in the case of CCICED and 2010 in the case of CAECC. In the 2012–2015 period of the incubation of the GSCI concept, these organisations made use of what has been termed "referent power," in the sense of leaders or institutions that provide informal recognition instead of formal reward. In the context of China GSCI, these organisations (i) represent a welcoming gesture to local governments and businesses that local initiatives will have high political profile at the national level producing reputational benefit, and (ii) ensure there will be a flexible alternative to the top-down "command and control" approach often embodied in campaigns or where formal legislative authority is sought. The two

Table 3 Central signals towards the four GSCI pilot cities

	MEP signal	Are there MIIT pilot companies for GSCM in the city	The key associated nonstate institutions
Tianjin	Nudging: the initial idea comes from central level, and the city responses with strong interest.	No	APEC commitment (linked to CAECC)
Shanghai		Yes	CCICED; CAECC
Shenzhen	Supporting: the initial idea comes from city level, and the central level responses with supporting signals, retrospectively.	Yes	–
Dongguan	Approval: the city applied for a proper title, and the central level approved.	No	CAECC

organisations created the initial momentum for the local initiatives and provided a certain level of reassurance regarding the "central signalling and observing" in Fig. 2. Even so, when MEP thought it had learned enough, it restored a more formal top-down approach in late 2015 by approving Dongguan's application of GSC pilot city. In MEP's statement of approval, MEP asked Dongguan to "*Extensively learn from the experience of Shanghai, Tianjin, Shenzhen and some leading enterprises to implement green supply chain management, to explore green supply chain labels, and to pilot the standard-making of green supply chain management system, and establish the performance evaluation system.*"

By contrast, MIIT has taken a more formal approach to promoting GSCIs. Green supply chain management has become an essential part of its green manufacturing scheme, together with national pilot projects for Green Factories, Green Products and Green Industrial Parks. MIIT decides on all the standards and metrics for evaluating and approving pilot companies. For each pilot company, a MIIT-verified, third-party evaluator is required to implement the evaluation process. The evaluators include research institutes, NPO consultancies as well as private consultancies, all of which are nonstate actors. Both Shanghai and Shenzhen have responded to MIIT's opportunity by acting as facilitators to encourage local companies to apply for MIIT GSC pilots.

Local Variations

Given the significant differences in the socio-economic context and the landscape of stakeholders, each city developed its own pathway to explore GSCIs. This subsection examines these pathways in terms of three dimensions: (i) who takes strong leadership? (ii) what are the strategies for implementation? and (iii) how have knowledge and technical support been provided? Table 4 summarises key findings.

Tianjin. The Tianjin government has published a series of policies, guidelines and standards to encourage green supply chain management. The initiative started in two industries (steel production and building & construction) where the supply chain is less complicated and the environmental impacts of suppliers are relatively simple. The GSC-related policies do not focus on creating higher environmental standards, but rather on the stage-setting process that highlights "*this is what we will do for our GSCI, and we shall do it right*" before embarking on bold steps. This approach reflects the influence of Deng Xiaoping's famous directive to "cross the river one stone at a time." This approach to the initial stage assures that key actors (both state and nonstate) have an essential role to play. Through its industry-specific approach, however, Tianjin has left limited space for the broader business community to contribute. For example, policymakers have designated only a few companies and organisations, mostly state-owned enterprises, to become the pilot cases.

Shanghai. The Shanghai Environmental Protection Bureau (EPB) decided to play a steering role and not invest heavily in new institutional arrangements. Thus, the rule-making process for GSCI in Shanghai adopted a public-private collaborative approach from the very beginning. In contrast to Tianjin's formal planning and policymaking at the

Table 4 Local exploration in the four GSCI pilot cities

	Leadership *Legitimacy or reward power*	Nonstate actors	
		Implementation and Facilitation *Referent power*	Knowledge and technical support *Expert power*
Tianjin	Strong institutional establishment across city departments;	Industrial associations and consultants as facilitator; Participating businesses are mainly SOEs	CEC (consultancy attached to MEP)
Shanghai	Shanghai EPB; Multinational companies	Voluntary actions of various industries and companies	Private Consultants; Reporting and disclosure platform
Shenzhen	Shenzhen Committee of settlement and environment; Leading domestic companies	Administrative guidance from local government Buyer-supplier collaboration	Private Consultants
Dongguan	Dongguan EPB; Department of Env. Protection, Guangdong province	Industrial association (furniture)	EDF (International NGO) CEC (consultancy attached to MEP) Reporting and disclosure platform

municipal level, Shanghai started by introducing three industrial standards in green supply chain management, which were direct outcomes of consultancy and pilot projects, respectively, with IKEA, General Motor and the Bailian Group (a Chinese giant in chain stores and retailing). Based on the three initial pilot projects, the GSCI of Shanghai scaled up into a more extensive programme called "Green Supply Chain 100+" to introduce best practices of green supply chain management to a broader community of both domestic and foreign companies. A Shanghai-based company can apply to join the "GSC 100+" programme and make a voluntary commitment to implement environmental management projects with at least three of its suppliers. Various teams of external expert from different nonstate actors have delivered most of the capacity-building activities to the participating companies. There are so far no direct financial resources allocated as a reward for participation in GSCI, but the city will issue a reputational award to encourage participation.

Shenzhen. In 2008, Shenzhen signed a voluntary agreement on green procurement with 24 companies, including both domestic and foreign companies. The government agreed to provide environmental performance information on 683 suppliers to the focal companies, and the focal companies agreed that they would not procure any products or services from any of the 683 suppliers with a record of environmental violations. This agreement produced a solid foundation for Shenzhen's GSCI starting in 2014. Similar to Shanghai, the GSCI of Shenzhen mainly works through encouraging focal companies and business leaders to participate. The local government plays only a "nudging" role. The Human Settlements and Environment Commission of Shenzhen,[11] which is the responsible key state actor in GSCI, has commented:

> The principles for (local) government to participate in green supply chains shall be the maximum of help and the minimum of interference. The government shall provide help regarding funding, technologies or information, to both buyer companies and supplier companies. In the meantime, the government shall not interfere with the decision-making process of companies, so that the green supply chain initiative becomes a market-driven process.

With such a mindset, Shenzhen's government applied an administrative guidance approach to facilitate GSCI at the local level. In general, the focal buyer companies are leading the initiative, supported by Shentai Environmental Tech, a private consultancy. For example, the digital hardware giant, Huawei, is the first company to participate in GSCI. Since the end of 2014, Huawei has established a Green Partnership scheme for its tier-1 suppliers. This is a private certification, which allows Huawei to select suppliers based on their environmental compliance, including hazardous materials, pollutant emissions and so on. Huawei's Green Partnership scheme and its efforts on GSCM have been recognised by the corporate information transparency index (CITI) ranking produced by the Institute of Public and Environment (IPE), whose work is the subject of another chapter in this volume. Those practices were developed before an intensive wave of central signals in Table 1. However, as Table 3 stated, the central government has retrospectively expressed strong support and recognised what has been tested—a gesture of saying "what you have been doing is appreciated" to Shenzhen.

Dongguan. Since the 2015 MEP approval, Dongguan has come up with a three-year action plan, including: Green public procurement; Green supplier development; Information platform of GSCM; Standard-making of GSCM; and Pilot projects in selected industries. A multi-stakeholder centre was co-founded by the Dongguan EPB, CAECC and the Environmental Defence Fund (EDF) to lead the work of GSCI. The most important outcome for the GSCI in Dongguan so far is the design and implementation of the Dongguan Index, a measurement and reporting system to evaluate the GSCM performance of name brand companies. In 2016, the Index was used to assess 17 furniture companies and nine shoe-making companies, since both industries are fundamental to Dongguan's economy. There are also plans to apply the Index to the electronics and machinery industries. Business participation in the Dongguan Index is not voluntary. Companies in those industries were told by local EPBs that they must provide essential data and thus cannot opt out. But those who gain a five or four-star rating in the Dongguan Index will be rewarded financially, in addition to achieving a reputational benefit.

Although each city has received a different format of central signals and has taken various strategies to explore GSCI practices, we think it would be very interesting and crucial to understand the "mechanism of influence" during such process. Traditionally, the legitimacy or reward power in Table 4 plays a vital, if not the sole role, in China's environmental initiatives. The effective of legitimacy power is based on comparable performance indicators, e.g. GDP or pollution control outcomes. Because of the complexity of GSC challenges, the central signals have used the central-level nonstate actors to create incentives for local actors to participate. Such local participation can lead to informal recognition instead of formal reward. Our assumption is that legitimacy power and comparable performance indicators create competition, i.e. all participants would follow a similar approach of environmental initiatives to gain political benefits. But referent power and informal recognition would likely create a pre-competitive environment for local actors to learn from each other and develop different strategies with a common vision—in this case, GSCM.

The Roles of Local Nonstate Actors in GSCI

We have discussed the advantages of having Beijing-based nonstate actors (e.g. CCICED and CAECC) provide a link between central government agencies (e.g. MEP or MIIT) and local actors. At the local level, local nonstate actors play different roles in different cities.

First, and most importantly, newly-founded organisations help to fulfil GSCI needs. These organisations play the role of *facilitators*. They have been designed to manage the local network of stakeholders interested in green supply chains, so that they can attract and encourage external organisations to participate in GSCIs. They are not state actors. But given the context of their "birth," they have referent power based on their relations with governmental departments as well as local companies, consultancies or other nonstate organisations. In effect, the state actors are endorsing the efforts of nonstate actors to create a mechanism of influence that differs from direct policy designing and implementation.

Empowered by the Tianjin municipal government, for example, two local nonstate actors have become the facilitators in implementing the governmental-led initiative. Both organisations have the status of *min ban fei qi ye* (non-governmental, non-profit unit). Their main tasks focus on policy implementation and GSCI corporate liaison. The Tianjin Green Supply Chain Association is a government created *shi ye dan wei* (governmental affiliated civil society) that facilitates the local rule-making process leading eventually to the development of green supply chain related standards and guidelines. The Tianjin Green Supply Chain Centre acts as a local consultancy (registered as a company but ultimately a non-profit organisation) to provide technical support for any companies who are willing to get involved (or become a local business leader) in GSCI. The multi-stakeholder GSCM centre in Dongguan, discussed above, operates much like the Tianjin Green Supply Chain Centre and has strong connections with Dongguan's EPB.

Second, there are both for-profit and non-profit organisations that act as *gatekeepers* of knowledge (Lazaric et al. 2008) in GSCIs. For example, the GSCI of Shanghai has established a supply chain reporting and disclosure platform. The platform is designed to engage in real-time data collection to track environmental compliance on the part of manufacturing companies in major cities and industrial parks of China. In the near future, the platform aims to provide a one-stop service for focal companies to monitor the environmental performance of their suppliers in order to mitigate supply chain risks and improve efficiency and resilience. Both Tianjin and Dongguan have plans to establish a similar platform. In the MIIT-led green supply chain pilot applications, there are dozens of MIIT-verified, third-party evaluators that provide knowledge and technical expertise to help the applicants understand GSCM and develop in-house GSCM systems. Those nonstate actors represent the *expert power* stated in Table 4. The mechanism of expert power aims to form coalitions and knowledge-sharing actors and hubs to contribute to solving sustainability challenges, and to create a pre-competitive environment for local state actors, i.e. include expert opinions when two cities are competing for a central recognition or award, instead of purely based on a political process.

Third, the different paths of city-level GSCIs can be categorised in terms of the relative role of state and nonstate actors. This distinction may be expressed in terms of institutions with formal legal authority vs. those with authority based on referent influence or valued expertise. Shanghai and Shenzhen, for example, developed their GSCIs through a public-private collaboration mechanism. Focal companies, whether foreign or domestic, have taken a significant leadership role in their initiatives, while the state actors are not heavily invested in regulatory efforts. This mechanism leads to increased inclusiveness and flexibility in contrast to the approach employed by Tianjin and Dongguan. However, it may be a bold approach, since it requires significant efforts to engage stakeholders and sufficient encouragement to make sure there is sufficient voluntary participation at the local level.

While there is significant involvement in GSCI on the part of nonstate actors, it remains rare to see engagement by community-level grass-roots actors. Why is this? There are several plausible hypotheses. Under Chinese regulations, with the exception of some government-created NGOs that have national reach (e.g. the All China Environmental Federation), NGOs are generally limited to action within the locality in which they are chartered; "cross-jurisdiction" action (i.e. coordination with NGOs chartered in another locale) is difficult and often politically sensitive. By definition, multinational supply chains cross local and global jurisdictional borders. Many suppliers involved in GSCIs are located in less developed Chinese cities or regions rather than in one of the four pilot cities. Civil society organisations in the four cities are likely to have less interest, knowledge, to reach out to those located elsewhere.

Conclusion

The cases analysed in this chapter show that China's strategy of using centrally-sponsored local experiments has already provided multiple paths for the design and implementation of local green supply chain initiatives. The central-local signals and responses in GSCIs are informal and have involved many nonstate actors at both levels, many of which have not been observed in other central-local relationships in China. Green supply chain management appears frequently as a keyword in central government policy documents promoting green manufacturing as well as environmental protection. Many actions have taken place, induced by central government signals without any need for formal legislation or direct orders coming down from MEP. National think tanks, such as CCICED and CAECC, have played an essential part during such nudging and signalling, but they do not provide rules for local GSCI. Their participation may be seen as a gesture to give local state and nonstate actors political confidence to proceed. In some cases, national think tanks can also provide technical expertise and links to GSCI-related resources.

It is too early at this stage to determine the impact of the GSCI efforts in the four pilot cities. But the hypothesis in this chapter is that central signalling can be both turbulent and ambiguous. For cities such as Shanghai and Shenzhen, central signalling can become less effective when a new round of political initiatives comes out. Shanghai and Tianjin had both become involved in the Zero-Waste Cities programme by the end of 2018; Shenzhen was awarded the status of "*Demonstration pilot zone for socialism with Chinese characteristics*" in August 2019. Although those changing signals may still be positive to green supply chain management, they create noise from the perspective of GSCIs, and somehow mitigate (or at least distract) the influence of key local nonstate actors. To avoid such "signal distraction," there is a need to pay more attention to specific GSCI initiatives that are NOT based on city-level business networks and associations.

Another hypothesis is that, the development of GSCIs may help the Chinese central government by mobilising resources in subnational regions to enhance governance capacity and to encourage businesses to regulate upper streams of supply chains located in areas of lower governance capacity both in China and elsewhere. Success in this effort would be a substantial contribution to the rectification of traditional "local protectionism" and cross-border regulation challenges.

Last but not least, GSCIs constitute experiments in a new and perhaps controversial strategy for private companies (both foreign and domestic) to collaborate with the municipal governments regarding environmental issues. GSCI provides a mechanism in which companies in China are not merely following the mandatory directives of central or local governments. Through GSCIs, companies are negotiating with local governments about what environmental issues to prioritise, and then helping state actors to achieve environmental goals in a mutually-beneficial way. This suggests that green supply chain management in China has economic and political implications extending beyond China's borders. These implications are highlighted and punctuated by the ongoing trade war between China and the United States. In the event that the trade war becomes more severe, Chinese investors will seek to outsource part of the manufacturing supply chain to other developing countries, so that final products imported into the United States will no longer be labelled "made in China." During such process of supply chain relocation, China sometimes needs to become a supply chain manager or coordinator instead of a supplier, and GSCIs may provide a practical approach to bargaining and collaborating with nonstate actors globally, to prepare China's industries to operate in a globalised setting, especially in the context of the Belt and Road Initiative. In taking an active part in rule-making processes for global supply chains,[12] China may improve the capacity and resilience of Chinese businesses operating overseas.

Notes

1. http://greatermekong.panda.org/challenges_in_the_greater_mekong/.
2. https://www.chinadialogue.net/article/show/single/en/10272-China-cleans-up-its-act-on-environmental-enforcement.
3. https://thediplomat.com/2017/12/china-cleans-up-its-act-on-environmental-enforcement/.
4. https://www.cnbc.com/2017/12/06/chinas-drive-for-cleaner-energy-is-causing-a-gas-shortage-for-winter.html.
5. http://www.worldstopexports.com/chinas-top-10-exports/.
6. https://e360.yale.edu/features/chinas_grand_plans_for_eco-cities_now_lie_abandoned.
7. https://thediplomat.com/2018/03/what-does-xi-jinpings-new-phrase-ecological-civilization-mean/.
8. https://www.ft.com/content/67a5e066-9faa-11dd-a3fa-000077b07658..
9. http://www.cciced.net/cciceden/.
10. https://www.cnbc.com/2018/09/14/china-must-do-more-to-tap-locals-in-belt-and-road-initiative-panel.html.
11. The name was effective from 2009 to 2019. Prior to 2009 it was Shenzhen Environmental Protection Bureau and after 2019 it changed the name back to Shenzhen Ecology and Environment Bureau, which echoes to the ministerial change at central level (Ministry of Ecology and Environment).
12. Opinions on promoting supply chain innovation and application, State Council, Oct 2017.

References

Arts, Bas. 2006. Nonstate Actors in Global Environmental Governance: New Arrangements Beyond the State. In *New Modes of Governance in the Global System*, ed. Mathias Koenig-Archibugi and Michael Zürn, 177–200. Springer.

Cashore, Benjamin William, Graeme Auld, and Deanna Newsom. 2004. *Governing through Markets: Forest Certification and the Emergence of Nonstate Authority*. New Haven: Yale University Press.

Dallas, Mark, Stefano Ponte, and Timothy Sturgeon. 2017. A Typology of Power in Global Value Chains. Copenhagen Business School, Working paper: 1–35.

Doh, Jonathan P., and Terrence R. Guay. 2006. Corporate Social Responsibility, Public Policy, and NGO Activism in Europe and the United States: An Institutional Stakeholder Perspective. *Journal of Management Studies* 43 (1): 47–73.

Handfield, Robert B., Steve V. Walton, Lisa K. Seegers, and Steven A. Melnyk. 1997. 'Green' Value Chain Practices in the Furniture Industry. *Journal of Operations Management* 15 (4): 293–315.

Karamichas, John. 2010. Environmental Governance in China. *Environmental Politics* 19 (3): 493–494.

Lazaric, Nathalie, Christian Longhi, and Catherine Thomas. 2008. Gatekeepers of Knowledge versus Platforms of Knowledge: From Potential to Realized Absorptive Capacity. *Regional Studies* 42 (6): 837–852.

Liu, Jianguo, and Jared Diamond. 2008. Revolutionizing China's Environmental Protection. *Science* 319 (5859): 37–38.

Liu, Lingxuan, Shotaro Matsuno, Bing Zhang, Beibei Liu, and Oran Young. 2013. Local Governance on Climate Mitigation: A Comparative Study of China and Japan. *Environment and Planning C: Government and Policy* 31 (3): 475–489.

Liu, Nicole Ning, Carlos Wing-Hung Lo, Xueyong Zhan, and Wei Wang. 2015. Campaign-Style Enforcement and Regulatory Compliance. *Public Administration Review* 75 (1): 85–95.

Mol, Arthur P.J., and Neil T. Carter. 2006. China's Environmental Governance in Transition. *Environmental Politics* 15 (02): 149–170.

Raven, Bertram H. 1993. The Bases of Power: Origins and Recent Developments. *Journal of Social Issues* 49 (4): 227–251.

Tanner, Murray Scot. 2005. Campaign-Style Policing in China and Its Critics. *Crime, Punishment, and Policing in China*: 171–188.

Teegen, Hildy, Jonathan P. Doh, and Sushil Vachani. 2004. The Importance of Nongovernmental Organizations (NGOs) in Global Governance and Value Creation: An International Business Research Agenda. *Journal of International Business Studies* 35 (6): 463–483.

Young, Oran R., Dan Guttman, Ye Qi, Kris Bachus, David Belis, Hongguang Cheng, Alvin Lin, Jeremy Schreifels, Sarah Van Eynde, and Yahua Wang. 2015. Institutionalized Governance Processes: Comparing Environmental Problem Solving in China and the United States. *Global Environmental Change* 31: 163–173.